The Hawai'i Garden
Tropical Exotics

The Hawai'i Garden
Tropical Exotics

HORACE F. CLAY and JAMES C. HUBBARD
photographs by RICK GOLT

University of Hawaii Press • Honolulu

First edition 1977
Paperback edition 1987

98 97 96 95 94 93   6 5 4 3 2

Library of Congress Cataloging in Publication Data

Clay, Horace Freestone, 1918–
  The Hawai'i garden.

  Includes index.
  CONTENTS: 1. Tropical exotics.—2. Tropical shrubs.
  1. Gardening—Hawaii—Collected works. 2. Plants,
Ornamental—Hawaii—Collected works. 3. Tropical
plants—Hawaii—Collected works. 4. Landscape gardening
—Hawaii—Collected works. 5. Plants, Cultivated—
Hawaii—Collected works. I. Hubbard, James C., joint
author. II. Golt, R. S. III. Title.
SB453.2.H3C55   635.9'09969   77-7363
ISBN 0-8248-0465-1 (v.1)
ISBN 0-8248-0466-X (v.2)

ISBN 0-8248-1127-5 (v.1 pbk)

The Hawai'i Garden volumes are dedicated to Elizabeth Loy McCandless Marks, who has worked selflessly and tirelessly to further the interests of tropical botany and horticulture in Hawai'i.

Tropical Exotics is dedicated to Laura Nott Dowsett, who has constantly supported ornamental horticulture and landscape gardening in the Islands.

Contents

Contents

Preface

Hawai'i is the home of one of the world's greatest collections of tropical and subtropical plants. Ever since the earliest Polynesian voyagers from the South Pacific brought their plants to these mid-ocean islands, already richly endowed with their own interesting native flora, untold thousands of plants from other tropical and temperate regions have come in a steady and mounting stream. Indeed, the Hawaiian Islands, with its wide range of benign and varied microclimates, have accepted plants from many different places—from humid jungle rain forests and arid deserts, and from seacoasts sprayed with salt and mountainsides of almost Andean heights.

Throughout the years, in addition to the numbers of plants that have been introduced, much hybridization has been accomplished in these islands, producing innumerable new varieties of hibiscus, crotons, anthuriums, orchids, colored ti, and many other plants. Hawai'i has led in the research and development of hybrids for production of varieties of pineapples and sugarcane that have been grown on an industrial scale. Hawai'i's renowned botanists—men like Joseph F. Rock, Harold St. John, and Otto Degener—have added immeasurably to the world's botanical knowledge of tropical plants.

Many residents of Honolulu during the nineteenth century, especially William Hillebrand and Don Francisco de Paula Marin, were zealous importers of new plants that added to the food supplies or enhanced the beauty of the Islands. Many of present-day Hawai'i's ornamental plants were found growing in the wild or in gardens in the world's tropical outposts by plant-loving travelers who sent or brought seeds, cuttings, or rootstocks to Hawai'i. Many rare and beautiful plants thus introduced were first grown in private gardens throughout the Islands, and in some instances they may still be seen in those very places. In truth, the Islands today are one great botanical garden, where—thanks to the efforts of passionate gardeners for two hundred years—perfect specimens of thousands of species gathered from the world around may be seen both commonly and easily.

Because most imported plants grow so easily and quickly somewhere in this encouraging subtropical land, new introductions are quickly propagated and dispersed. Often, a plant that is rare in one season may be found in abundance within a year or so. Our residents' generosity and willingness to share assures newly introduced plants a rapid and wide dissemination.

With the enormous variety of plants that surround us wherever we go comes also a great curiosity and search for knowledge about them. Everywhere people ask, when they see a new one they like, "What plant is that?" "Where will it grow?" "How do I care for it?" It is our hope to help the plant lover and amateur gardener to know, use, and care for the beautiful, unusual, or merely interesting plants that are at his disposal in modern Hawai'i.

Tropical Exotics and its companion volume, *Tropical Shrubs,* are intended to serve as a guide for all Hawai'i's gardeners to the cultivation and enjoyment of the Islands' ornamental plants. Detailed information is given about each of the species chosen as outstanding examples of plants in specific categories: the species history and the origin of its name; how it grows, where it grows best, and how it is best used in the garden; how to propagate, prune, and fertilize it; and, also important, any serious disadvantages it may have as a garden plant. In order to make the information about each individual species as complete as possible, some repetition is unavoidable, particularly in discussions of closely related plants. To minimize such repetition, appendixes are added covering material that is generally applicable to the plants presented in the volume—detailed information on insect pests and plant diseases and their control, and plant propagation techniques.

Occasionally in the text the name of a plant or plant group will be seen printed in bold face type; this indicates that the plants thus set apart are more fully described elsewhere in the volume.

A few words of explanation about the choice of photographs: The intention has been to depict each plant, not in the conventional, literal, "whole plant" style, but in an interpretative, artistic manner. For the most part, the photographs concentrate on a small physical area of the subject, creating an image that is visually pleasing, and always true to the nature of the plant and expressive of its particular individuality.

The arrangement of plant families in these volumes follows the system used by A. Engler and K. Prantl, German botanists, who placed them according to complexity, beginning with the simplest of plants and ending with the most complex. Plant species within the families are arranged alphabetically according to their Latin binomials.

Botanical nomenclature is constantly—and necessarily—being revised, and, just as constantly, authorities disagree. For these books, we have accepted *Willis' Dictionary of the Flowering Plants and Ferns,* edited by A. Shaw (1973) as authority for family and generic names. The species and varietal names in this volume have been authenticated by Hawai'i eminent botanist, Dr. Harold St. John. The authors are deeply grateful for his painstaking work.

Much of our information about the history and practical uses of these plants was derived from three works: Marie C. Neal, *In Gardens of Hawaii* (1965); J. C. Th. Uphof, *Dictionary of Economic Plants,* 2nd ed. (1968); and I. H. Burkill, *A Dictionary of the Economic Products of the Malay Peninsula,* 2nd ed. (1966).

Meanings and origins of plant names were obtained primarily from *The Royal Horticultural Society Dictionary of Gardening,* ed. F. J. Chittenden, (1956). Also helpful were R. S. Woods, *The Naturalist's Lexicon* (1944); L. H. Bailey, *How Plants Get Their Names* (1933); *Webster's New Collegiate Dictionary* (1976); and *The Encyclopedia Britannica,* 15th ed.

In addition to these sources and to *Willis' Dictionary* (already mentioned), we have consulted many other references: L. H. Bailey, *The Standard Cyclopedia of Horticulture* (1947), and *Manual of Cultivated Plants* (1949); *Curtis's Botanical Magazine* (1787–1975); J. Bateman, *The Orchidaceae of Mexico and Guatemala* (1843, reprinted 1974); M. A. Reinikka, *A History of the Orchid*

(1972); F. Sander, *Reichenbachia* (1888–1894); V. Padilla, *The Bromeliads* (1973); R. Austin and K. Ueda, *Bamboo* (1970); A. Engler and K. Prantl, *Die natürlichen Pfanzenfamilien* (1887–1914); A. Graf, *Exotica 3* (1976); *Index Londonensis* (1929 and supplements); *Index Kewensis* (1895 and supplements); *Paxton's Magazine of Botany* (1834–1849); P. C. Standley, *Trees and Shrubs of Mexico* (1920–1926); H. O'Gorman, *Mexican Flowering Trees and Plants* (1961); L. E. Bishop, *Honolulu Botanic Gardens Inventory, 1972* (1973); H. St. John, *List of Flowering Plants in Hawaii* (1973); R. B. Streets, *The Diagnosis of Plant Diseases* (1969); C. L. Metcalf and W. P. Flint, *Destructive and Useful Insects,* 4th ed. (1962); H. T. Hartmann and D. E. Kester, *Plant Propagation: Principles and Practices,* 3rd ed. (1975); *Flora Pacifica* catalog (1970); Staff of L. H. Bailey Hortorium, Cornell University, *Hortus Third* (1976).

We wish to express our deep appreciation for the generous financial support and unstinting cooperation given by the Stanley Smith Horticultural Trust, the Bradley L. Geist and Victoria S. Geist Foundation, Elizabeth Loy McCandless Marks, Ellen Doubleday, Laura Nott Dowsett, Dorothy Bading Lanquist, and Mr. and Mrs. Alfred J. Ostheimer III.

Our thanks go also to many others: Mr. John Gregg Allerton, Mr. and Mrs. J. Garner Anthony, Dr. John W. Beardsley, Mrs. John H. Beaumont, Dr. and Mrs. Adrian Brash, Mrs. Lillian Craig, Dr. Gilbert S. Daniels, Dr. and Mrs. Thomas F. Fujiwara, Mr. John K. Hayasaka, Dr. Derral Herbst, Mr. and Mrs. Oscar M. Kirsch, Mr. and Mrs. Tamotsu Kubota, Mr. Pat Kawamoto, Mr. J. L. Merkel, Dr. Wallace C. Mitchell, Mr. and Mrs. W. W. Goodale Moir, Mr. Herbert C. Shipman, Dr. Harold St. John, Mr. Hiroshi Tagami and Mr. Richard Hart, Dr. and Mrs. Richard K. Tam, Mrs. H. Alexander Walker, Mr. and Mrs. J. Milton Warne, Mr. Paul R. Weissich, Mr. and Mrs. Warren Q. K. Yee, and Mr. and Mrs. Iwao Yokooji.

For assistance of many kinds we are indebted also to the Friends of Foster Garden, the Harold L. Lyon Arboretum, the Honolulu Academy of Arts, the Honolulu Botanic Gardens, the Garden Club of Honolulu, the Outdoor Circle, the Bernice P. Bishop Museum library, the Hawaii State Library, and the libraries of the University of Hawaii.

A Word of Caution

Do Not Eat or Taste Parts of Unfamiliar Plants

Plants have been used since time beyond measure both as foods and as medicines. In many cases, the plants are benign agents, producing little or no ill effect when they are eaten—and often they are credited with doing much good. Many plants, however, are very irritating to the body's tissues, internally or externally, and in some cases can cause severe sickness or death. The bits of information about uses as medicines and foods offered in these books are included only because of their historic interest. This mention must not be construed as being a recommendation for their use as foods or medicines. Often, highly poisonous plants must be thoroughly processed before their toxic properties are removed, making them safe to use. Failure to prepare plants properly may lead quickly to illness or death.

The general rule, when encountering unknown plants is this: ***Do not eat or taste any part of any unfamiliar plant.*** Many of Hawai'i's garden plants, including some of those listed in these books, contain irritating or poisonous agents that may be highly detrimental to the user. Never apply parts of unknown plants to the skin or eyes or mouth, and never eat them. Many age-old remedies based upon tropical plants are beneficial, but often their worthwhile attributes are based upon how and when and in what state of preparation they are used. The uninformed person should never try these remedies.

Modern scientists are studying ancient folk remedies to determine whether beneficial medicines can be obtained from them. Until the plants are tested adequately, however, the home medicine chest and the family menu should not include any unfamiliar plants.

The Hawai'i Garden
Tropical Exotics

Pandanaceae
(Screw Pine Family)

The screw pine family includes three genera: *Pandanus, Freycinetia,* and *Sararanga.* About 700 species comprise the family, at least 600 of which are placed in the genus *Pandanus.* The long and durable leaves of screw pines, such as the hala we know in Hawai'i *(Pandanus odoratissimus),* have been used since before people first ventured into the Pacific islands, for making shelters, mats, clothes, baskets, sails, navigational charts, and many other things. The fruit, rich in vitamin C, is eaten by many Pacific islanders, either fresh or dried. The dried fruits helped to sustain seafarers on long ocean voyages, as well as settled folk during periods of famine. In Hawai'i, the best-known member of the family other than hala is the vine 'ie'ie *(Freycinetia arborea).* Hawaiians plaited the aerial roots of the 'ie'ie into a strong network foundation for helmets, feather capes, baskets, and fishing gear.

The generic term *Pandanus* is derived from the Malayan word *pandang,* a dialect name for several species local to that region. Early European explorers called the whole group screw pines, because of the spiral arrangement of the leaves on branches, and because both the fruits and the foliage are very similar in appearance to those of the **pineapple *(Ananas comosus).*** In fact, at one time some plants now classified as *Pandanus* were considered to be species of *Ananas.*

Pandanus veitchii
Variegated Screw Pine, Variegated Hala

The variegated screw pine is seen quite often as an ornamental in Hawaiian gardens. It is not a native of Hawai'i, but rather of southern Polynesia. Its more useful close relative, *Pandanus odoratissimus,* which Hawaiians called hala, is native to Hawai'i as well as to many islands in central and southern Polynesia. Today, more than 600 species of *Pandanus* are recognized; most are native to tropical regions stretching from Africa to Indonesia.

The variegated screw pine looks strikingly "tropical" in a landscape. Its trunks and their stiltlike supporting roots, topped with giant tufts of green and white swordlike leaves, are very dramatic, especially in places bright with sunshine or night-lights. Generally it is treated as a specimen plant, although sometimes it is seen in groves.

Pandanus is the Latinized version of the Malayan name, *pandang,* used throughout Malaysia for the several native species of screw pines; *veitchii* honors the celebrated English nurserymen James Veitch, Jr. (1815–1869), and his son John Gould Veitch (1839–1870).

HABIT
A spreading, umbrella-shaped, evergreen tree that grows to about 25 feet in height. Mature plants produce many stiltlike aerial roots along the trunk and lower branches which serve as props to support the heavy crown. Foliage heads are composed of many leathery, sword-shaped leaves, variegated in green and white, each about 4 feet in length, with either smooth or toothed edges. Plants are either male or female; females produce the interesting pineapple-shaped fruits; males exhibit large flower clusters with creamy white bracts. Fruits, deeply divided, yield large, wedge-shaped "keys" that become orange upon ripening. Slow growth rate; quite easily transplanted at any size.

GROWING CONDITIONS
Very adaptable; will grow in almost any location, although this species does prefer hot, dry beach conditions. Like most kinds of *Pandanus,* it grows happily in the teeth of salt-laden winds. It will not withstand extreme drought.

USE
Specimen plant; mass planting; container plant; colorful foliage.

PROPAGATION
Large branches may be removed from the parent plant and set out directly in a new garden location, where they will root readily. Or separate the individual "keys" from the fruit cluster, and lay them flat on their sides, with the top face showing above the sand or potting soil. Seeds will germinate within 8 to 12 weeks.

INSECTS/DISEASES
To control mealybugs, apply diazinon or malathion.

PRUNING
Remove dead and badly damaged branches only at the nearest crotch because, once cut, a branch will not grow any more but will remain a dead stump.

FERTILIZING
Apply general garden fertilizer (10-30-10) to the planting bed at 4-month intervals, and to container plants at monthly intervals.

DISADVANTAGES
Constant dropping of leaves and fruits.

Gramineae
(Grass Family)

This family is one of the largest in the plant kingdom: at present it consists of about 620 genera and more than 10,000 species. Grasses of one kind or another have been found in almost every part of the world where plants can grow, some in extreme Arctic or Alpine conditions, others in the hottest, most arid of deserts. Generally, the grasses are sunlovers; most species grow in nature's open spaces.

Many of the world's land-bound animals would be unable to exist without the grasses. Many insects, birds, and mammals depend upon grasses as their main, sometimes their only, source of food. And man would certainly be the poorer without corn, wheat, barley, oats, millet, rice, rye, sorghum, bamboo, and sugarcane. Aside from their value as foods, many grasses are useful in other ways. It is probable that the bamboos, those giant grasses, have been the most widely used of all plants for shelter, utensils, and other aids in daily life. The bamboos' strength, suppleness, versatility, and availability have led people to devise innumerable applications for leaves, stalks, shoots, and roots.

The family name Gramineae comes from *gramen,* a Latin word for grass in general.

Arundinaria nitida
Black-stemmed Bamboo

Black-stemmed bamboo originated in the cool, temperate Kansu and Szechuan regions of China. Nonetheless, it is one of the bamboos that can flourish through a rather wide range of temperatures, for it is also quite at home in the warm tropical regions of the world. The Chinese use the handsome dark canes for making heavy-duty implements and in construction. Basketware, such as sieves and carriers, are plaited from pliable lengths split from the shiny black stems.

This bamboo makes a handsome garden plant. It is moderately large, growing to more than 20 feet in height in its native habitat. Hawai'i's form is shorter, rarely exceeding 12 to 15 feet. The canes are somewhat more slender than those of other bamboos, and are easily recognized by their shiny, black-green color. As with most bamboos, this species requires a fairly large garden area in which to grow. It soon sends out underground shoots that become the new canes and expands rapidly into rather large clumps. It is not unusual to find a new cane shooting up several feet away from the central cluster.

Arundinaria comes from *arundo,* a Latin word for cane; *nitida,* meaning having a smooth and polished surface, refers to the stems. The genus *Arundinaria* contains about 150 species.

HABIT
An erect, shrubby, evergreen bamboo that grows to about 15 feet in height; densely clumping, spreading by means of underground stems. Small, distinctively colored purple-green-black canes grow to be about ½ inch in cross-section. Exhibits typical bamboo foliage of long, narrow, pointed, grasslike leaves. In the manner of most bamboos, the clump flowers only after having grown for several decades, sets seed, then dies. Moderate growth rate; easily transplanted.

GROWING CONDITIONS
Quite adaptable but requires constant ground moisture; grows well in full sun or partial shade. Will not grow well in beach areas, where salt winds burn the foliage.

USE
Specimen plant; hedge; mass planting; container plant; colorful canes.

PROPAGATION
Easily propagated by root division.

INSECTS/DISEASES
For scale, apply malathion.

PRUNING
Remove dead and damaged canes; this species may be pruned into formal hedge shapes; when used in this way, canes are completely hidden by foliage.

FERTILIZING
Apply general garden fertilizer (10-30-10) to the planting bed at 3-month intervals, and to container plants at monthly intervals.

DISADVANTAGES
Unless root mass is contained, plant may spread into areas where not wanted. Starts taken from very old plantings may be short-lived inasmuch as all segments of one bamboo clump will flower, set seed, and die simultaneously.

Bambusa angulata
Chinese Dwarf Bamboo, Square-stemmed Bamboo

This very attractive native of China is quite distinctive in that its dark green stems grow naturally in squared, slender, flat-sided columns. In the Orient, Chinese dwarf bamboo often is used as decorative material, especially in light construction and paneling, screening, and garden fencing. Its squared form makes it easily adaptable to architectural detailing. Many relatives among bamboos also are widely grown for practical purposes. *Bambusa beecheyana* and *B. spinosa* are primary sources of edible bamboo shoots. When growing plants for those shoots, the Chinese heap soil over the bases of the bamboo clumps, so that they are protected from the light. At harvest time the soil is carefully removed, and the tender shoots thus exposed are cut close to ground level. *B. arundinacea, B. balcooa,* and *B. polymorpha* yield thicker and sturdier canes that are useful in heavier construction. Fishing poles oftentimes are made from the willowy *B. multiplex* or *B. nana.*

Although *B. angulata* is not often seen in Hawaiian gardens, there is no reason why it should not be more popular. It is not especially aggressive or dense. Individual canes stand apart a bit from each other, giving much the same visual effect as that of bamboos painted on an Oriental screen.

Bambusa—and the English word bamboo—derive from bambu, the Malay name for these gigantic grasses; *angulata,* meaning having angles, describes the squared canes of this species. The Chinese name for bamboo is chu.

HABIT
This is a relatively short species of bamboo, growing to less than 15 feet in height. Its loose, airy habit, with canes well spaced, allows easy visibility through the clump; distinctively squared, dark olive-green, matt-textured canes are from ½ to ¾ inch in cross-section; downward-curving, clawlike "thorns" encircle each joint. Bright green, feathery leaves are irregularly interspersed throughout the clump. Like all bamboos, this one will flower after many years of growth and set seed, whereupon the whole clump dies. Though individual canes grow rapidly, the clump increases slowly in size. Easily transplanted at any size.

GROWING CONDITIONS
Very easily grown except in exposed beach locations. Prefers cool, very moist, protected areas with soil rich in humus. Grows well in full sun or partial shade.

USE
Small bamboo grove; container plant; distinctively squared canes; cut sections often are used as flower containers.

PROPAGATION
Easily propagated by root division.

INSECTS/DISEASES
To control scale, spray with malathion.

PRUNING
Remove old or damaged canes only. In general, healthy canes need to be removed only to open up the mass of the clump.

FERTILIZING
Apply general garden fertilizer (10-30-10) to the planting bed at 6-month intervals and to container plants at monthly intervals.

DISADVANTAGES
None; this is not an aggressive species.

Bambusa vulgaris var. *aureo-variegata*
Golden-stemmed Bamboo

One of the most common green-stemmed bamboos is *Bambusa vulgaris,* parent of this most colorful and attractive variegated offspring. The parent, one of the largest of the grasses, grows to a height of 50 feet and more in its native region, which stretches from Indonesia to India. It is one of the chief sources of the pulp used in making fine papers for letter writing and elegant calligraphy. Shoots of *B. vulgaris* are eaten throughout Asia. The strong, light, long, supple stalks are highly valued in the construction trades. *B. vulgaris* is considered by many Orientals to have medicinal value. During the Middle Ages in Europe, alchemists produced a potion called *tabachir* from bamboo canes as an antidote for poisons. Among several contemporary medical uses is one favored by the Javanese, who prescribe water accumulated within the golden bamboo tubes as a treatment for jaundice.

This golden-stemmed species is probably Hawai'i's most often used ornamental bamboo. It was an early introduction into Hawai'i, and possibly may have arrived here before Captain Cook's time. The earliest Polynesian settlers in these islands are credited with having introduced 'ohe (as they called bamboo), but whether this species and variety were among the kinds they brought is open to conjecture.

Bambusa derives from the Malayan name, bambu; *vulgaris,* meaning common, refers to the green-stemmed parent's ubiquitous presence in its native region; *aureo-variegata* means variably colored with golden yellow.

HABIT
A very large, woody, evergreen grass that grows to 50 feet and more in height; easily recognized by its colorful canes, striped in green and yellow; mature stalks generally reach 4 to 5 inches in diameter. Graceful, narrow, papery, bright green leaves festoon the slender willowy branches at the extreme top of each cane. Individual shoots grow rapidly, quickly attaining mature height. As with all bamboos, after several decades the entire grove flowers, sets seed, then dies. Easily transplanted.

GROWING CONDITIONS
Very adaptable; will grow in most Hawaiian locations, but prefers cool, moist, protected areas with soil rich in humus. Grows well in full sunlight or partial shade.

USE
Specimen clump; bamboo grove; container plant; colorful canes; edible shoots.

PROPAGATION
Easily propagated by root division. Or, cut canes into 4- to 5-foot sections, each containing two or more nodes and half bury them horizontally in ground; both roots and leaf shoots will develop at the nodes.

INSECTS/DISEASES
For scale, use malathion.

PRUNING
Remove old or damaged canes only. This species is not of a densely clumping nature, and does not require thinning.

FERTILIZING
Apply general garden fertilizer (10-30-10) to the planting bed at 6-month intervals, and to container plants at monthly intervals.

DISADVANTAGES
Clumps may grow to be too large for the average garden unless they are restrained.

10

Cortaderia selloana
Pampas Grass, Cortadera

Pampas grass, as its name indicates, grows profusely in the Argentine pampas, the vast plains that stretch across the central part of that country. Spanish-speaking pioneers named the razor-sharp grass *cortadera,* which means "cutting readily" or "easily cut." Gauchos—and their mounts—wear protective covering made of leather when riding through pampas grass. Leaves of this grass are used in preparation of a good commercial paper, and the roots are said to be of some medicinal value.

Pampas grass is one of the best known of the many hundreds of dried materials employed in decorative arrangements. It was introduced early into North American gardens and residences for decoration. Nineteenth-century parlors almost always exhibited a sheaf of feathery pampas grass flower heads set in a china vase. Although less popular today, it is still used decoratively in contemporary homes. In the garden it is generally grown as a specimen for the sake of its fountain of foliage, topped with feathery plumes—a light, airy, delicately gray-green contrast to the usual garden colors.

Cortaderia is a Latinized version of the Spanish word, *cortadera,* which describes the extremely sharp leaf edges; *selloana* is named for Friedrich Sello (1789–1831), a German botanical collector who explored in Brazil.

HABIT A high, mounding, evergreen grass that grows in dense clumps 6 feet or more in height; individual blades, about ½ inch wide, are gray-green in color, reach a length of 6 feet or more, and rise vertically from the clump's base and arch delicately downward, the outermost blades' tips nearly touching the ground. The leaf surfaces and edges are armed with tiny sawteeth, making each leaf a lacerating hazard. Slender flower stems rise straight upward and produce handsome feathery-white flower heads that seem to float above the foliage; flower heads may grow to 8 feet or more in height. In Hawai'i the main flowering period is from spring to late fall. Moderate growth rate; small root divisions are easily transplanted.

GROWING CONDITIONS Highly adaptable; will grow almost anywhere in Hawai'i; will withstand temperatures much colder than found in the usual Hawaiian climates. Like most grasses, it requires a regular water supply to maintain healthy green foliage; drought conditions will quickly dry portions of the foliage, causing a brown, unkempt appearance. Grows best in areas of full sun, but will tolerate partial shade.

USE Specimen clump; container plant; distinctive foliar color; barrier planting.

PROPAGATION So easily propagated by root division that seeds are generally not used.

INSECTS/DISEASES Not subject to insect infestations in Hawai'i.

FERTILIZING Apply general garden fertilizer (10-30-10) to the planting bed at 4-month intervals, and to container plants at monthly intervals.

DISADVANTAGES The leaves' sharp edges can lacerate the skin quickly and painfully.

Cyclanthaceae
(Cyclanthus Family)

The cyclanthus family, from tropical America, is moderately small, having 11 genera and about 180 species. To the casual observer, it would appear to be a part of the much larger palm family in that all its members exhibit palm-like foliage. Botanists, however, put them into a group of their own. Not a few botanists consider this family to be the link between the palms and the arums **(Araceae)** because the cyclanthus species, palmlike in foliage, bear flowers like those of the arums.

A few species are found in tropical gardens; the two described in this book probably are the best known for landscaping purposes. Three species have economic value in some countries. Both *Carludovica angustifolia* from Peru and *C. labela* from Mexico produce leaves that are used for thatching. *C. divergens,* from Brazil and Peru, is the source of a fiber that Peruvians make into cordage. A fourth species, **C. palmata,** produces the fiber used in making Panama hats.

The family name Cyclanthaceae is derived from *Cyclanthus,* one of the family's eleven genera. This term, compounded from two Greek words, *kuklos,* meaning circle, and *anthos,* meaning flower, describes the circular arrangement of the flowers upon the rodlike spadix.

Carludovica palmata
Panama Hat Palm, Palmita, Toquilla

The Panama hat palm is not a palm. The hats made from this plant—popularly thought to be produced in Panama—are manufactured chiefly in Ecuador and, to a lesser degree, in Colombia and Peru. The plant itself is native to Peru. Women of the region harvest the unopened leaves, then make a "straw" suitable for plaiting by boiling the leaves, cutting them into thin strips, and then drying these in bright sunshine. Six leaves can provide enough material for an average hat. Latin Americans consider the Panama hat to be essential for summer wear, whether or not hats are in fashion elsewhere. Material not good enough to be made into hats is plaited into handsome baskets and matting. Brooms are made from the coarsest material.

This plant is quintessentially tropical. Great, unblemished, finely pleated, fan-shaped leaves rise on rich green stalks to form a rooflike canopy. The plant is exhibited effectively in shaded garden nooks, small courtyard spaces, and protected "jungle" walkways. In season, the brilliant orange and red flower clusters seem to emit sparks of color in deeply shaded garden areas.

Carludovica is named for both King Charles IV of Spain (1748–1819) and his Queen, Luisa (1751–1819); *palmata,* meaning palmlike, describes the appearance of the plant. Central Americans know it as palmita (little palm) or toquilla (bonnet, cap).

HABIT
An evergreen plant of moderate size that grows to 15 feet or more in height. Highly decorative, pleated, fan-shaped leaves, 3 feet in diameter, form at the end of tall slender canelike stems that emerge in dense clusters from trunks near the ground. Decidedly clumping in habit. In the spring, foot-long spadices emerge from the foliage base, gradually mature, then split open to display succulent red and yellow pulp containing many small white seeds. Moderate growth rate; successful transplanting requires much care.

GROWING CONDITIONS
This tropical jungle plant demands a great deal of protection. Grows best in moist, shaded, wind-free areas, in humus-rich, well-watered, well-drained soils. Not suitable for beach or dryland garden conditions.

USE
Specimen plant; container plant; tropical foliage and colorful flower clusters.

PROPAGATION
Almost always propagated by root division. To propagate from seeds, spread pulp-covered seeds over moistened tree fern fibers and keep constantly moist.

INSECTS/DISEASES
Relatively insect- and disease-free in Hawai'i. Treat occasional scale infestation with malathion. For occasional mealybug infestation, use diazinon or malathion.

PRUNING
Remove old and damaged leaves and stems at trunk.

FERTILIZING
Apply general garden fertilizer (10-30-10) to the planting bed at 4-month intervals, and to container plants at monthly intervals. Add generous amounts of humus to heavy clay soils before planting, and periodically thereafter as a top dressing.

DISADVANTAGES
Leaves will burn in exposed situations.

Cyclanthus bipartitus
Cyclanthus, Portorrico, Hoja de Lapa

This plant, native to the jungles of Central and South America and the Caribbean islands, is used in some parts of those regions as thatching and for its weavable fibers. Close relatives in the genus *Carludovica* are used for the same purposes. The plant's Spanish names are portorrico (native of Puerto Rico) and hoja de lapa (leaves like cleavers).

It comes from a jungle habitat. In nature it is found growing on humid forest floors, completely screened from searing winds and burning sunlight. It is distinctive in appearance and character; its fluted V-shaped leaves immediately distinguish it from any other plant. Stiffly vertical, scissors-shaped leaves rise above garden ferns and other kinds of groundcovers in delicate, slender-stemmed clumps. Slight breezes catch the tall, saillike leaves, starting gentle undulations within the clumps.

Cyclanthus bipartitus is the only species in the genus. The generic name, compounded from *kuklos,* meaning circle, and *anthos,* meaning flower, describes the circular arrangement of flowers along the spiky flower spadix; *bipartitus,* meaning divided into two, refers to the parted leaves that are joined at the base.

HABIT	An evergreen plant of moderate size that grows 12 to 15 feet in height. Extremely unusual V-shaped leaves, about 3 feet long and 12 inches wide, appear to be two separate leaves joined at the stem. Near the ground, slender leaf stems emerge in clusters from short trunks. Clumping in habit. During the spring blooming period, tall green flowers, somewhat like those of anthuriums, emerge from the central trunks. Slow growth rate; successfully transplanted only with much care.
GROWING CONDITIONS	Must be constantly shaded, protected from sun, wind, and mechanical damage. Requires considerable and constant moisture.
USE	Specimen plant; large container plant; lush tropical foliage.
PROPAGATION	Almost always propagated by root division. May be started from seeds, which are spread over moistened, crushed tree fern fibers and kept constantly moist.
INSECTS/DISEASES	Relatively insect- and disease-free in Hawai‘i. Treat occasional scale infestation with malathion. For occasional mealybug infestation, use diazinon or malathion.
PRUNING	Remove old and damaged leaves and stems at trunk.
FERTILIZING	Apply general garden fertilizer (10-30-10) to the planting bed at 4-month intervals. Add generous amounts of humus to heavy clay soils before planting and periodically thereafter as a top dressing.
DISADVANTAGES	Leaves will burn in exposed situations.

Araceae
(Arum Family)

The arums are a rather large family, numbering about 115 genera and 2,000 species, most of which are native to the world's tropics. A representative of the small minority from the temperate zones is the calla lily *(Zantedeschia aethiopica)*. The term calla lily is really a misnomer, because the plant is an arum, not a lily (see **Liliaceae**). The two arums of most importance are taro *(Colocasia esculenta)*, whose roots and greens are a staple food of Pacific island peoples, and yautia *(Xanthosoma sagittifolium)*, whose secondary roots are prepared and consumed by tropical Americans in much the same way as Pacific islanders use taro. Most arums produce an irritating sap; edible kinds are made palatable by a variety of cooking processes which dissipate the irritating components.

The great Greek scientist Theophrastus (third century B.C.), the "father of botany," used the term *aron*, when describing an arum-like plant in his historic work, *Inquiry into Plants*. Later botanists appropriated this term for the family Araceae. Nowadays the terms "arum" and "aroid" (resembling arum) are used interchangeably.

Aglaonema commutatum
Aglaonema, Chinese Evergreen

This aglaonema is one of 20 or so species all of which are native to the moist, jungle regions of Indomalaysia. Quite a number of these have been introduced into ornamental horticulture, especially *Aglaonema commutatum, A. marantifolium, A. oblongifolium,* and *A. pseudobracteata. A. commutatum* probably is the species most often planted in Hawai'i.

In Malaysia this plant grows wild in the cooler, moist uplands on many of the islands and on the mainland peninsula. Because of the plant's toxicity, in some places in Malaysia the roots are processed into a worming tonic.

Aglaonemas are somewhat sturdy plants that grow well in protected, partly shaded locations. Often they are used as container plants for house and lanai. In fact, the aglaonemas are among the few plants that can grow healthily for long periods of time in poorly lighted and badly ventilated rooms. Aglaonemas closely resemble their near relatives, the **dieffenbachias**.

Aglaonema, from *aglaos,* meaning bright, and *nema,* meaning thread, refers to the glistening flower stamens; *commutatum,* meaning changed or altered, describes the variegated foliage. On the mainland, aglaonemas introduced into the nursery trade as houseplants commonly are called Chinese evergreen, but they are known by their generic name in Hawai'i.

HABIT | An erect, herbaceous, evergreen shrub that grows to about 4 feet in height; compact crowns of leaves flourish at tops of bare, vertical stems; the large, leathery, green leaves mottled with white are about 8 to 10 inches long. Insignificant flowers appear in small clusters, then give way to clusters of glossy, bright red berries; flowering and fruiting occur periodically throughout the year. Moderate growth rate; easily transplanted.

GROWING CONDITIONS | Prefers cool, moist, protected areas, rich soil, partial or full shade; ideally suited for Hawai'i's moist upper valleys and forest landscapes. Not a good beach or dryland plant, where it will be exposed to salt-laden or drying winds.

USE | Specimen plant; mass planting; container plant; house plant; tropical foliage and colorful berries.

PROPAGATION | Leafless stem cuttings are readily rooted in well-drained potting soil or in garden beds. To propagate from seed, remove seeds from red coating and plant immediately in moist, well-drained potting soil. New foliage forms are produced from seeds.

INSECTS/DISEASES | For thrips, apply diazinon or malathion. For scale and mealybugs, use malathion.

PRUNING | Remove dead and mechanically damaged leaves. To improve old and unsightly plants, remove entire crown at desired stem height; new growth will appear at the top of the cut stump. The cut crown will also root, to produce another plant.

FERTILIZING | Apply general garden fertilizer (10-30-10) to the planting bed at 3-month intervals, and to container plants at monthly intervals.

DISADVANTAGES | Calcium oxalate crystals present in leaves and stems may cause severe irritation of throat and mouth, if these plant parts are eaten raw.

Alocasia macrorrhiza
Variegated 'Ape, Elephant Ear, Ta'amu, Biga

This interesting foliage plant is a horticultural form of Polynesia's 'ape. The vivid green-leafed parent, *Alocasia macrorrhiza,* a native of Sri Lanka (Ceylon), India, and the Malay Peninsula, has been a food crop in Indomalaysia and southern Polynesia since people first went there to live. The rhizomes of 'ape produce an important staple starch in those areas of tropical Asia and Polynesia where rice is not preferred. Sometimes 'ape stems are cooked in coconut milk, then baked in underground ovens, to make a most tasty dish.

Variegated 'ape, although of similar nutritive value, generally is planted for its beauty. Foliage forms abound. All the 'apes, whether variegated in foliage or not, introduce large, bold, tropical effects into a garden. They are mammoth arums that can easily dwarf a man.

Alocasia is a variant of *colocasia,* a Greek word taken from the Arabic *kolkas* or *kulkas,* that was associated with the lotus root *(Nelumbium speciosum); macrorrhiza,* meaning large root, describes the starchy, edible, rhizome. In Hawai'i, *A. macrorrhiza* has several common names, for example, 'ape (Hawaiian), elephant ear, ta'amu (Samoan), and biga (Filipino).

HABIT
A very large, herbaceous evergreen plant that grows to 15 feet or more in height. Great succulent leaf stems jut vertically from a large, short trunk. A huge, 4- to 5-foot, arrowhead-shaped, ruffled leaf tops each stem. Occasional calla-like flowers rise from the trunk, enclosing attractive red berries within the spadix; at ripening, the spadix peels away to expose the fruits. Moderate growth rate; easily transplanted.

GROWING CONDITIONS
Quite adaptable if grown in protected forest conditions. Not a plant for beach or dry ridge locations. Prefers cool, moist places and soils rich in humus. Grows well in either shade or sun provided that considerable humidity and complete protection against wind are constantly available.

USE
Specimen plant; mass planting; large container plant. Although the sweet-scented flowers are of some topical interest, the plant is planted solely for its gigantic leaves.

PROPAGATION
New offshoots that sprout from a parent at ground level may be set out in new garden locations. The plant may also be propagated by cutting large trunks into short pieces and half-burying each section on its side in soil. Also may be propagated from seeds, which may produce new foliage forms.

INSECTS/DISEASES
For thrips, apply diazinon or malathion.

PRUNING
Remove dead and damaged leaves only. Drastic, complete, pruning at the trunk will result in vigorous but somewhat slow regeneration of the crown. 'Apes are much like tree ferns in manner of growth: as old leaves die, new ones emerge to replace them.

FERTILIZING
Apply general garden fertilizer (10-30-10) to the planting bed at 3-month intervals, and to container plants at monthly intervals.

DISADVANTAGES
Although it is a well-known food plant, *alocasia* should never be eaten without adequate preparation, including cooking, by an experienced person.

Anthurium andraeanum
Anthurium, Capotillo Colorado

This most popular of all anthuriums is one of about 550 *Anthurium* species known. The group is native to tropical America and the Caribbean islands.

Most anthuriums are primarily ornamental, but *Anthurium oxycarpon* has leaves that when dried release a musky vanilla scent. These leaves, called *folha cheirosa* (fabulous leaf) in Portuguese, are mixed as an aromatic ingredient in snuffs and fancy smoking tobaccos. Ornamental anthurium flowers are used often in Latin America on special occasions.

The flowers are very popular because of their brilliant colors and durability even after they are cut. One new flower sprouts with each leaf; healthy plants should produce about eight flowers during the year. Massed plants provide a year-round, colorful, natural, yet exotic cover. This species was brought to Hawai'i from England by S. M. Damon in 1889.

Anthurium, from *anthos,* meaning flower, and *oura,* meaning tail, refers to the numerous tiny true flowers that form on the yellow taillike spadix; *andraeanum* is named for Edouard F. André (1840–1911), botanist and horticultural editor in Europe. A. *andraeanum* was discovered for botany in the wilds of Colombia by José J. Triana, who sent specimens to André in 1876. Spanish-speaking peoples know it as capotillo colorado (little red cape).

HABIT A small, herbaceous, evergreen plant that usually grows to about 3 feet in height. Shiny, foot-long, leathery, heart-shaped leaves unfurl from tops of very slender rod-like stems. Older plants send down thin aerial support roots. Colorful capelike spathes frame the taillike spadices; the spathes may vary in color from green to orange to pink to red to white, depending on the horticultural variety. Ripening spadices produce seeds. Slow growth rate; easily transplanted.

GROWING CONDITIONS Must be grown in shaded garden or lathhouse situations in moist but well-drained planting mixtures. Plants grow well in combinations of soil, bark, sawdust, tree fern fiber, macadamia nut shells, coconut husks, charcoal, crushed rock, or cinders.

USE Specimen plant; mass planting; container plant; colorful tropical flowers and foliage.

PROPAGATION Established horticultural varieties are reproduced from offshoots from the parent plant. New horticultural varieties and color forms are developed from seeds; spread seeds over moistened finely crushed tree fern fiber and keep constantly moist.

INSECTS/DISEASES For thrips, apply diazinon or malathion. For scale and mealybugs, use malathion. For anthracnose disease of the spadices, spray captan fungicide on the flower heads, using 1 teaspoon of fungicide per gallon of water. During dry weather, spider mites may deform foliage and flowers; control with malathion.

PRUNING Remove dead and damaged leaves and flowers only.

FERTILIZING Apply time-release 14-14-14 pellets to the planting bed or container at 4-month intervals; water regularly.

DISADVANTAGES None of any consequence in Hawai'i.

26

Anthurium andraeanum cv. 'Hawaiian Butterfly'

Obake Anthurium

The obake anthurium hybrids are entirely Hawaiian in origin. Local commercial growers have used improved varieties of the tropical American species **Anthurium andraeanum** as parent stock to develop very different and unusual variegated anthuriums. Obake hybridization began in the early 1930s.

The common name, obake, is a Japanese word for ghost. It alludes to the extraordinary differences among hybrids in form, size, and color exhibited by the waxy spathes. The Hawaiian Butterfly obake, a delicate combination of green, pink, and yellow, originated in the Hilo area and is so named because the triangular spathes resemble butterfly wings.

In the garden, obake anthuriums are grown in quite the same way as are their red-flowered ancestors. They require considerable shade and protection, but they bring interesting form and added color to sun-shy areas. They are excellent as cut flowers. Arrangements of them last two to three weeks.

Anthurium, from *anthos,* flower, and *oura,* tail, refers to the tiny true flowers on the taillike spadix; *andreanum* is named for Edouard F. André (1840–1911), botanist and horticultural editor in Europe.

HABIT
A small, herbaceous, evergreen plant that grows to about 3 feet in height. Shiny, foot-long, leathery leaves unfurl from tops of very slender rodlike stems. Older plants send down thin aerial roots to support the foliage mass. Highly variegated capelike spathes in combinations of green and pink surround a bright yellow spadix. (Other obakes show different form and color combinations.) Ripening spadices produce seeds. When well grown, plants produce year-round blooms, generally about eight flowers and leaves per year. Slow growth rate; easily transplanted.

GROWING CONDITIONS
Must be grown in shaded garden or lathhouse situations in specially prepared, moist but well-drained planting mixtures. Each anthurium expert relies on his own mix but in general, plants grow well in combinations of soil, bark, sawdust, tree fern fiber, macádamia husks, charcoal, and construction-grade crushed rock or cinders.

USE
Specimen plant; mass planting; container plant; colorful tropical flowers.

PROPAGATION
Established horticultural varieties are reproduced from offshoots from the parent plant. New horticultural varieties are developed from seeds; spread seeds over moistened, finely crushed tree fern fiber and keep constantly moist. Sometimes older plant tops are removed with a few aerial roots and replanted.

INSECTS/DISEASES
For thrips, apply diazinon or malathion. For scale and mealybugs, use malathion. For anthracnose disease of the spadices, spray captan fungicide on the flower heads, using 1 teaspoon fungicide per gallon of water. During dry weather spider mites may deform foliage and flowers; control with malathion.

PRUNING
Remove dead and damaged leaves and flowers only. If plants become leggy and unsightly, remove entire top with some aerial roots attached and replant in place.

FERTILIZING
Apply time-release 14-14-14 fertilizer pellets at 4-month intervals.

DISADVANTAGES
None of any consequence in Hawai'i.

Anthurium crystallinum
Crystal Anthurium

This anthurium is prized for its foliage, not for its flowers. It is native to the jungles of Colombia and Peru, where its glistening leaves reflect shafts of sunlight filtering through the forest canopy. Within the last few decades the plant has become popular, entering the hothouse and nursery trades of more temperate climes. The crystal anthurium, very tropical in habit and appearance, is one of the most beautiful of the *Anthurium* species grown in ornamental horticulture. As with most of the other 550 or so related species, this anthurium probably has little, if any, medical or culinary use.

A forest-floor plant, the crystal anthurium is most advantageously displayed as a single specimen or arranged in masses beneath great trees. Garden viewers, inspecting the large handsome leaves, will discover transparent surfaces under which faceted tissues catch and reflect the sun's rays.

Anthurium, from *anthos,* meaning flower, and *oura,* meaning tail, refers to the numerous small flowers that form on the taillike spadices; *crystallinum,* meaning crystalline, describes the distinctive reflecting surfaces of the leaves.

HABIT
A small, herbaceous, evergreen plant that grows to about 3 feet in height. Foot-long, dark green, dramatic, velvety leaves are laced with strong white raylike veins. The plant generally grows from decaying vegetable matter deposited on the forest floor; older plants, as their trunks grow in height, send down aerial roots to support the otherwise top-heavy mass. The plant has interesting but unprepossessing flowers that are vaguely reminiscent of commercial anthurium flowers. As with other anthuriums, viable seeds develop along the flower spadices. Moderate growth rate; easily transplanted.

GROWING CONDITIONS
A forest-floor dweller, the plant does best in protected, shaded, moist areas, in soils or soil mixes containing large amounts of humus. Tree bark, crushed tree fern fiber, macadamia nut husks, sawdust, charcoal, and cinders added in generous amounts to the soil, provide excellent substrates. The plant will not grow well in beach or ridge gardens where it would be exposed to salt, wind, or sun.

USE
Specimen plant; mass planting; container plant; dramatic, tropical foliage.

PROPAGATION
Usually propagated from seeds; spread seeds over moistened, finely shredded tree fern fiber and keep constantly moist. Or remove top with some aerial roots, and set it out in a new location.

INSECTS/DISEASES
For scale and mealybugs, apply malathion. For thrips, use diazinon or malathion.

PRUNING
Remove dead and damaged leaves and old flower stems only. This is essentially a foliage plant; often gardeners will remove the flowers to increase its beauty.

FERTILIZING
Apply time-release 14-14-14 fertilizer pellets to the planting bed or container at 4-month intervals.

DISADVANTAGES
None of any consequence in Hawai'i.

Anthurium hookeri
Bird's-nest Anthurium

The bird's-nest anthurium is native to the Caribbean islands and the Guiana region of South America. A forest-floor inhabitant, the plant grows in stiff-leafed rosettes that may reach a diameter of 8 feet or more; such breadth distinguishes the species as one of the largest of all the anthuriums. Hawai'i's gardeners have called it the bird's-nest anthurium because it resembles the native tree-dwelling bird's-nest fern *(Asplenium nidus)*.

This is a large-scale garden plant, creating a spectacular display, even when grouped with other highly dramatic leaf and flower forms. Being a shade lover, it is best planted under an arboreal canopy where only moderate amounts of sunlight may reach the garden floor. Although its leaf spread may be extensive, the actual stem and root system of this anthurium is very compact. For this reason, it is easily grown in containers, in which it can produce impressive whorls of leaves for the decoration of protected terraces and large lanais.

Anthurium, from *anthos,* meaning flower, and *oura,* meaning tail, describes the taillike spadices characteristic of the genus; *hookeri* commemorates Sir William Jackson Hooker (1785–1865), eminent botanist and director of Kew Gardens, London.

HABIT
A large, herbaceous, evergreen plant that grows to 6 feet or more in height and 8 feet or more in diameter. Large, leathery, spatulate leaves, each of which can be 6 feet long and 2 feet wide, sprout in a rosette arrangement from a low trunk; leaves have strong central midribs and secondary veins; leaf edges are characteristically wavy. Occasionally an unattractive purplish anthurium flower rises on a tall stem from the central trunk; sometimes purple fruits develop along the spadix. Moderate growth rate; easily transplanted at any size.

GROWING CONDITIONS
Requires partially shaded areas providing soils rich in humus, constant moisture, and protection from winds or falling debris. Optimum growth occurs in humid, almost hothouse conditions.

USE
Specimen plant; mass planting; large container plant; dramatic, tropical foliage; grown for leaves, not for flowers.

PROPAGATION
Generally propagated from seeds; spread seeds on the surface of moistened, finely shredded tree fern fiber and keep constantly moist.

INSECTS/DISEASES
For thrips, apply diazinon or malathion.

PRUNING
Remove flowers and badly damaged or dead leaves only; removal of healthy foliage produces a lopsided plant. Many gardeners who do not wish to propagate the plant from seeds remove the flowers as they appear.

FERTILIZING
Apply time-release 14-14-14 fertilizer pellets to the planting bed or container at 4-month intervals.

DISADVANTAGES
Handsome large leaves are easily damaged by excessive sun, wind, or falling debris. Raw plant parts may be toxic.

Anthurium scherzerianum
Pigtail Anthurium

This plant, which appears to be a horticultural oddity, is a true botanical species, naturally developed in the wilds of Guatemala and Costa Rica. Although the flowers resemble those of its more popular cousin, **Anthurium andraeanum,** the pigtail spadices that characterize *A. scherzerianum* are distinctly different in size and form. As a plant, it is somewhat smaller in size, and its leaves are much less glossy, as well as much more elongated and pointed than are those of the commercial anthurium.

Pigtail anthuriums are easily adapted to garden environments and are certain to be conversation pieces when grown as specimen plants. Grown in masses under large shade trees or in a jungle setting, they assume a more natural appearance.

Anthurium, from *anthos,* meaning flower, and *oura,* meaning tail, describe the plant's taillike spadices. *A. scherzerianum* was discovered in Guatemala by M. Scherzer (1821–1903), a Viennese plant collector.

HABIT — A small, herbaceous, evergreen plant that can grow to about 3 feet in height. Matt-textured, dark green, elongated, heart-shaped leaves unfurl at tops of slender stems. The root mass grows well in forest floor litter, decaying chunks of wood, bark, or fiber; older plants produce aerial roots which support the otherwise heavy crowns. Red, pink, salmon, or white capelike spathes contrast with distinctive curled, bright yellow spadices. Several horticultural forms have been developed from seeds (which ripen along the spadices). Plants, when well grown, produce blooms constantly. Slow growth rate; easily transplanted.

GROWING CONDITIONS — Like many anthuriums, this species must be grown in shaded gardens or lathhouses. Plants are best set in moist but well-drained, coarse soils containing liberal amounts of humus and vegetable fiber, bark, nut hulls, charcoal, crushed rock, or cinders.

USE — Specimen plant; mass planting; container plant; colorful and unusual flowers.

PROPAGATION — Established horticultural varieties are reproduced from offshoots from the base of parent stocks. Sometimes older plant tops are removed with a few aerial roots and replanted. New horticultural varieties and color forms are developed from seeds: spread seeds over moistened, finely crushed tree fern fiber; keep constantly moist.

INSECTS/DISEASES — For thrips control, apply diazinon or malathion. For scale and mealybugs, use malathion. For anthracnose disease of the spadices, apply captan fungicide, using 1 teaspoon fungicide per gallon of water. During dry weather spider mites may deform foliage and flowers; control with malathion.

PRUNING — Remove dead and damaged leaves and flowers only. If plants become top heavy and leggy, remove entire top with some aerial roots attached and replant in place.

FERTILIZING — Apply time-release 14-14-14 fertilizer pellets to the planting bed or container at 4-month intervals.

DISADVANTAGES — Raw plant parts may be toxic.

34

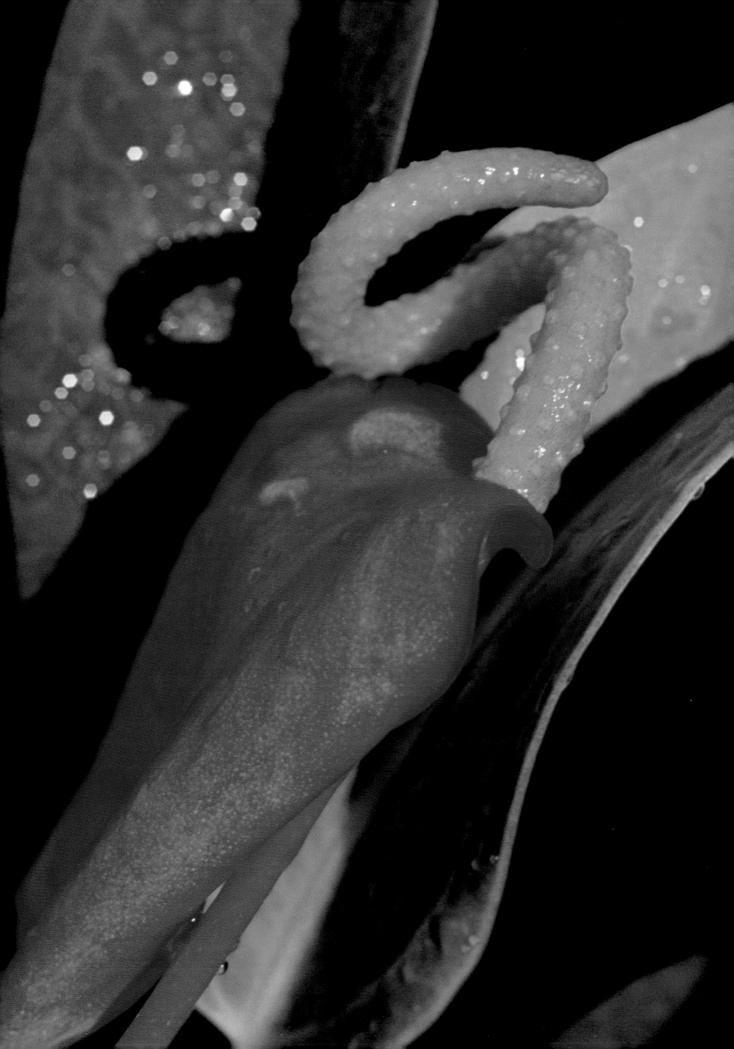

Anthurium sp.
Dwarf Purple Anthurium

New species of plants are still being discovered in nature and introduced to botany and ornamental horticulture. A case in point is the very attractive dwarf purple anthurium. Dr. Robert Dressler, a biologist with the Smithsonian Tropical Research Institute, Panama Canal Zone, collected the first wild specimens in the forests of Coclé del Norte, Panama, in 1972. Introduced into Hawai'i in 1973, this new discovery is being propagated locally for early presentation to Hawai'i's ornamental plant trade. The plant has not yet been described in botanical terms.

In its Panama home this anthurium is an epiphyte, growing naturally on tree limbs and fallen materials. Petite, perfect, purple flowers rise from a rosette of small strap-shaped leaves. Besides producing a flower of such unusual color, the plant also emits a slight fragrance suggesting eucalyptus oil. Its future in the art of landscaping depends upon the ease and speed of its reproduction. For some time, however, it will be encountered only as a rarity in Hawai'i—a collector's item.

Anthurium, from *anthos,* meaning flower, and *oura,* meaning tail, refers to the taillike spadices characteristic of the genus.

HABIT A small, herbaceous, evergreen plant whose leaves can grow to a length of about 4 inches. The narrow, strap-shaped leaves form in a tidy rosette; the plant's roots attach themselves to tree limbs or logs in a manner similar to that of epiphytic orchids. Slender flower stems rise from the leaf cluster to a height of about 8 inches. Distinctive purple flowers grow to about 1½ inches in overall size. Moderate growth rate; easily transplanted.

GROWING CONDITIONS Naturally an epiphytic tree-dweller; so far, the cultivated plant has been grown to best advantage in containers filled with 100 percent fir bark. Requires highly protected, shaded, moist atmospheric conditions for optimum growth. Basically a plant for the lathhouse or protected lanai garden.

USE Specimen plant; container plant; colorful flowers.

PROPAGATION Currently being propagated by offshoots; ultimately, cross-pollination should prove feasible.

INSECTS/DISEASES For thrips control, apply diazinon or malathion. For scale and mealybugs, use malathion. For anthracnose disease of the spadices apply captan fungicide, using 1 teaspoon fungicide per gallon of water. During dry weather spider mites may deform foliage and flowers; control with malathion.

PRUNING Remove dead flowers and leaves only.

FERTILIZING Apply time-release 14-14-14 fertilizer pellets to the planting medium at 4-month intervals. Plants growing epiphytically on trees should be helped every 2 months with liquid forms of commercial anthurium fertilizer.

DISADVANTAGES The plant is rare, and no doubt will remain so for a few years.

Caladium bicolor
Caladium, Kalo Kalakoa, Corazon de Maria

Caladium bicolor is one of 15 species in the genus, all of which are native to tropical South America and the Caribbean region. A horticultural variety, *C. bicolor* var. *poecile,* known in the West Indies as taioba brava (wild taro), produces rhizomes which, when boiled, are eaten like potatoes. Another horticultural variety, *C. bicolor* var. *vellozianum,* or mangara, has similar rhizomes that are edible when cooked and when fresh are used medicinally as an emetic or purge. Two Brazilian species, *C. sororium* and *C. striatipes,* have fruits and tubers, respectively, which are eaten by Brazilian natives.

C. bicolor was an early introduction to Europe from the New World. First transported to Madeira by early Portuguese voyagers, it found its way to England in 1773. Eighteenth-century Europeans made it a popular hothouse plant, and propagators avidly developed many different forms with variegated foliage, beginning a practice that continues to this day.

Caladium is the Latinized version of a Malay name, *kaladi,* which probably applies to a relative in Southeast Asia; *bicolor,* meaning two colors, describes the leaves of the original importation from America. Hawaiians call the plant kalo kalakoa (calico taro). It is called corazon de Maria (Mary's heart) in the Philippines, where it has escaped from gardens and grows wild.

HABIT
A small, deciduous, herbaceous plant that can grow to a height of about 3 feet. Multiple stems sprout from fleshy tuberous roots; each stem produces a colorful foot-long, heart-shaped leaf. Leaf shapes and colors vary greatly according to horticultural variety. Leaves fade and die down during the spring months. New foliage appears in the summer. The first leaves are green, but subsequent foliage unfolds in its variegated form. Insignificant calla-like flowers bloom during the growing season. Fast growth rate; easily transplanted during the dormant period in spring.

GROWING CONDITIONS
Very adaptable; prefers cool, moist garden areas of humus-rich soils. Caladiums must have adequate protection from wind, or damage to leaves will affect the plant's beauty. Most colorful growth is achieved in sunny or slightly shaded areas.

USE
Specimen plant; mass planting; container plant; grown for its colorful leaves.

PROPAGATION
Established horticultural varieties are propagated by root division; new varieties and color forms are developed from seeds.

INSECTS/DISEASES
For root mealybug control, drench soil around plant with malathion solution; or remove tubers from soil, dip roots in malathion solution, and replant in fresh, sterilized soil mixture.

PRUNING
During the growing season, remove dead and damaged leaves and flowers only; in spring, as the plant begins to fade, all leaves may be cut back to the ground.

FERTILIZING
Apply general garden fertilizer (10-30-10) to the planting bed at 3-month intervals, and to container plants at monthly intervals. Plants should be fertilized between the time when the leaves die back in spring and new growth appears.

DISADVANTAGES
Yearly dieback may leave unsightly voids in the garden. Raw plant parts are toxic.

38

Colocasia sp.
Variegated Colocasia

This native of Nicaragua is an ornamental relative of the Pacific island taros. The taros are all cultivars of *Colocasia esculenta,* which has been an important food crop in the world's tropics since people first discovered it. Thousands of agricultural varieties have evolved during the millennia since then. The early Hawaiians rank among the most successful developers of the plant as a foodstuff; during the several centuries after they arrived in these islands, they developed several hundred indigenous cultivars. Most of the Pacific islanders who eat taro boil or bake the tubers and eat them in the same manner that Westerners eat potatoes. But Hawaiians, after cooking the starchy root, also pound it into a smooth paste that they call poi. Many cultivars developed in Hawai'i are still grown ornamentally and commercially in Island gardens, either in dryland plantings or in irrigated paddies.

Variegated colocasia, a newcomer to Hawaiian ornamental horticulture was found in the Nicaraguan jungle in 1975. It is grown for its brilliant, dramatic foliage, not for food or flowers.

Colocasia is derived from old Arabic plant names, *kolkas* or *kulkas,* used in reference to the root of the lotus, *Nelumbium speciosum,* probably because plants of both genera yield edible roots. This plant, being relatively new to botany, has not yet been given a species name.

HABIT An evergreen, herbaceous plant that grows to about 18 inches in height. Very delicate and slender stems rise from the root to unfurl elegant, varicolored, arrowhead-shaped leaves; the distinctively pointed and elongated leaves grow to about 8 inches in breadth and 12 inches in length. Occasional and insignificant calla-like flowers sprout from the leaf clusters. Moderate growth rate; easily transplanted.

GROWING CONDITIONS Grows best in protected, shaded garden areas, in humid atmospheric conditions. Must be planted in moist but well-drained soils with high content of peat moss, leaf mold, composted bark, etc. Tender in structure, the plant requires more than the usual amount of protection from adverse influences.

USE Specimen plant; mass planting; container plant; grown for its exotic foliage.

PROPAGATION Propagated by dividing established plants at the roots.

INSECTS/DISEASES Like the taros, this plant is susceptible to attack by leaf hoppers; control by applying diazinon.

PRUNING Remove old flowers and damaged and dead leaves only. Many gardeners find the flowers somewhat unattractive and remove them at an early stage. Complete rejuvenation of entire plant can be achieved by cutting the crown to the ground. This drastic action generally is coordinated with the dividing of the tubers.

FERTILIZING Apply general garden fertilizer (10-30-10) to the planting bed at 6-month intervals, and to container plants at monthly intervals.

DISADVANTAGES Raw plant parts are toxic.

Dieffenbachia picta cv. 'Rudolph Roehrs'
Dumb Cane, Canna de Imbé

Although this and several others of the 30-odd species of *Dieffenbachia* are called dumb cane, the plant which was originally accorded that common name is *D. seguine* from the Caribbean. In old days, parts of this plant, in which the needle-shaped crystals of calcium oxalate are especially plentiful, were forced into the mouths of slaves to torture them, and literally to make them speechless. People of the Caribbean region have used the plant also as a counterirritant to assuage serious and deep-seated pains. The specially prepared medicine is applied to hurting joints, where it causes a somewhat irritating sensation, thus partly overwhelming that of the deeper pain—a treatment similar in principle to the use of applying a mustard plaster.

D. picta cv. 'Rudolph Roehrs' is a natural mutant that was discovered growing in the Roehrs Nursery in New Jersey in 1936. Forty years later this handsome foliar form is one of the most popular of all hothouse and tropical plants. The *Dieffenbachia*s will grow in most Hawaiian garden locations. With adequate protection from sun and wind, these green and white plants will introduce striking tropical effects in the garden or on the lanai.

Dieffenbachia is named for J. F. Dieffenbach (1794–1847), a German physician and botanist; *picta,* meaning painted, alludes to the interesting patterns of the leaves.

HABIT
A herbaceous, evergreen plant that grows to about 8 feet in height. Bare, looping and curved, hoselike stems support a thick crown of long oval leaves, variegated in green and white. Leaves may reach 2 feet in length. Insignificant calla-like flowers are borne periodically within the leaf cluster throughout the year. Normally the plant does not produce seeds in Hawai'i. Moderate growth rate; easily transplanted.

GROWING CONDITIONS
Very adaptable, will grow almost anywhere in Hawai'i if given protection from full sun and heavy wind. One of the most popular of container and house plants; will grow indoors over a long period of time. Grows best in rich, well-watered, well-drained soils in areas of filtered sunlight.

USE
Specimen plant; mass planting; container plant; tropical foliage.

PROPAGATION
Cut stem sections, with or without foliage, to any length and set upright in permanent location. Or half bury horizontal stem sections in planting mixture and keep moist.

INSECTS/DISEASES
For thrips and mealybugs, apply diazinon or malathion. For scale, use malathion.

PRUNING
Remove dead and damaged foliage. Some gardeners dislike the flowers, and remove them at an early stage.

FERTILIZING
Apply general garden fertilizer (10-30-10) to the planting bed at 2-month intervals, and to container plants at monthly intervals.

DISADVANTAGES
Calcium oxalate crystals, present in all *Dieffenbachia*s, will cause painful and debilitating mouth and throat irritations if parts of the plant are eaten raw.

Homalomena rubescens
Homalomena, Gandubikachu

Homalomena rubescens, known in India as gandubikachu, is native to the hill regions of East Pakistan, India, and Burma, especially near the Himalayas. It is one of about 140 species of *Homalomena,* many of which are native to Indomalaysia, and the remainder, indigenous to South America.

Characteristically, members of the genus possess aromatic roots. In India and Malaysia the roots of several species are used medicinally as counterirritants, either in the form of a poultice or as infusions for relieving mucus-clogged throats. Sometimes Malays will use a preparation of *Homalomena rubescens* roots as an aid to fishing. The poison, called *ipoh,* is cast into streams, where it stupefies the fish.

H. rubescens is a sprightly garden plant, rich in leaf texture and color. Its leaves are clear and smooth, looking as if they have been well scrubbed. New foliage, quite reddish at first, turns a deep green at maturity on the upper surfaces; lower surfaces and stems retain a sumptuous burgundy cast.

Homalomena, from *homalos,* meaning flat, and *nema,* meaning thread, describes the plant's flattened flower stamens; *rubescens,* meaning turning red, refers to the reddish hue of young leaves and of stems.

HABIT
A herbaceous, evergreen plant that grows to about 4 feet in height. Tall, fleshy, wine-red stems spring in clusters of two or three from the ground when plants are young, and higher, on fleshy stems, when plants are older. Clean, heart-shaped leaves, 10 inches wide and 12 to 14 inches long, unfurl at tops of stems. Mature leaves exhibit dark green coloring on upper surfaces; undersides are purplish red. Brilliant red, boat-shaped flower buds appear occasionally at different times throughout the year. Seeds are borne along the spadices of the calla-like flowers. Moderate growth rate; easily transplanted.

GROWING CONDITIONS
The cultivated plant grows best under jungle-like conditions. Must be protected from full sun and heavy winds. Grows best in rich, well-watered, well-drained soil. Debris falling from overhead branches can damage otherwise perfect foliage.

USE
Specimen plant; mass planting; container plant; colorful tropical foliage, stems, and flower bracts.

PROPAGATION
Propagated by root division. Sometimes older plants are removed with a few aerial roots, then replanted. Or, cut rootstalk into segments and half bury them horizontally in planting mixture.

INSECTS/DISEASES
For thrips and mealybugs, apply diazinon or malathion.

PRUNING
Remove dead and damaged leaves and flowers. An older plant may be rejuvenated by removing the leaf cluster, with some stem attached, and planting it in a new location. The remaining stump will send out new growth.

FERTILIZING
Apply general garden fertilizer (10-30-10) to the planting bed at 2-month intervals, and to container plants at monthly intervals.

DISADVANTAGES
Raw plant parts may be toxic.

Philodendron × cv. 'Multicolor'
Red-leafed Philodendron

This brilliant-leafed plant is a horticultural hybrid, the progeny of two of the 275 species of *Philodendron* known from tropical America and the Caribbean. Although this hybrid was developed for ornamental purposes, several relatives are used in a variety of ways in their native regions. Several possess medicinal properties which are employed in the treatment of skin diseases, rheumatism, and intestinal worms. The stems of *P. oxycardium* and the aerial roots of *P. sagittifolium* are used in making baskets. Both *P. selloum,* a common landscape plant in Hawai'i, and *P. bipinnatifidum* have edible fruits that, when properly harvested, can be eaten fresh or made into jellies.

The red-leafed philodendron has been introduced only recently into Hawai'i from Florida, where it was first developed in the 1960s by Robert McColley of Orlando's Bamboo Nurseries. It is one of the most colorful and attractive of all the exotic tropical plants. Its vivid multicolored foliage, caught in the sun's rays, seems to capture the brilliance of stained-glass windows. Because its parents are jungle plants, this cultivar, too, grows best in tree-shaded gardens.

Philodendron, from *philos,* meaning beloved and *dendron,* meaning tree, indicates the usual arboreal habit of most members in this genus. This cultivar is known as 'Multicolor' because its foliage develops in several color combinations, generally in mixtures of brilliant reds, greens, creams, and whites.

HABIT A herbaceous, evergreen plant that grows to about 5 feet in height. More shrubby than vinelike, the plant produces large, smooth, brilliant red, foot-long leaves in a tight cluster at the top of a heavy, fleshy stem. As plant grows in height, aerial roots are formed to steady the top-heavy crown. Very slow growth rate; easily transplanted.

GROWING CONDITIONS Semiepiphytic in nature, the plant grows best in substrates composed primarily of such materials as peat moss, chopped tree fern fiber, leaf mold, and artificial components such as perlite. It must be provided with partial shade, and its perfect foliage must be protected from falling debris. Requires constant moisture and well-watered, well-drained soil.

USE Specimen plant; mass planting; container plant; colorful foliage.

PROPAGATION Easily propagated by stem cuttings.

INSECTS/DISEASES For thrips and mealybugs, apply diazinon or malathion. To control scale, use malathion.

PRUNING Remove dead and damaged leaves and flowers.

FERTILIZING Apply time-release 14-14-14 fertilizer pellets to the planting medium at 4-month intervals.

DISADVANTAGES Raw plant parts may be toxic.

Philodendron wendlandii
Self-heading Philodendron

Philodendron wendlandii was first collected from the wild by Hermann Wendland, a noted nineteenth-century German botanist. He found the plants growing in Costa Rica and in Panama. Of the 275 or so *Philodendron* species known (all of which are natives of tropical America), about one hundred are grown ornamentally in Hawai'i. Most of these are climbing vines; however, the self-heading philodendron grows close to the ground in a well-ordered rosette. In appearance it is somewhat similar to the epiphytic bird's-nest fern, *Asplenium nidus*.

The plant is lush and succulent. Its translucent leaves and rotund petioles, when cut, all but spout the water stored in their tissues. Very much contained, the foliage grows in a spokelike arrangement, with each leaf rigidly set apart from the others. Because its natural home is the forest floor, this plant may be grown in protected locations, such as a lanai or a shaded garden. Its compact form also makes it suitable for growth in large containers.

Philodendron, from *philos,* meaning beloved, and *dendron,* meaning tree, describes the climbing arboreal habit of most species in the genus; *wendlandii* honors the species' discoverer, Hermann Wendland (1825–1903), director of the Royal Garden at Herrenhausen, Hanover, Germany.

HABIT	A herbaceous, evergreen plant that grows to about 3 feet in height. Large, glossy, foot-long, paddle-shaped leaves with prominent midribs, grow from fleshy, cylindrical stems that sprout from a very short, thick trunk. Insignificant, ivory-colored, calla-like flowers appear occasionally throughout the year. Slow growth rate; easily transplanted.
GROWING CONDITIONS	Prefers jungle conditions and rich, moist but well-drained soil. Best growth is achieved in places protected from excessive sun, wind, and falling debris.
USE	Specimen plant; mass planting; container plant; tropical foliage.
PROPAGATION	Most easily grown from seeds: remove ripe but undried seeds from flower spadix and spread immediately on moist, finely chopped tree fern fiber; keep constantly but moderately moist; seeds will sprout readily within a period of a few weeks. When well rooted, set out in containers filled with rich soil.
INSECTS/DISEASES	To control thrips, apply diazinon or malathion. For scale and mealybugs, use malathion. For spider mites, spray with wettable sulfur.
PRUNING	Remove only dead and damaged leaves and flowers; excessive pruning of healthy leaves will produce a lopsided plant. Some gardeners dislike the flowers and remove them as they appear.
FERTILIZING	Apply time-release 14-14-14 fertilizer pellets to the planting bed or container at 4-month intervals.
DISADVANTAGES	Raw plant parts may be toxic.

Spathicarpa sagittifolia
Spathicarpa

This diminutive aroid grows in moderately large numbers on the hot and humid jungle floors of tropical South America. It can be found chiefly in the states of Bahia and Rio Grande do Sul in Brazil, in Paraguay's Serra De Maracaju, and in the Guaraipo and Saltiño regions of the Argentine. This is one of seven species attributed to the genus *Spathicarpa.* The plant is grown chiefly for its attractive foliage and flowers, and has no medicinal or economic value.

Low-growing in character, spathicarpa appears to be a small-scale version of its larger aroid relatives, the **spathiphyllums.** Planting spathicarpa in great numbers provides a crisp green and white groundcover. It is also a charming potted plant, growing handsomely by itself or in combination with other tropical foliage forms. Because of its size, it lends itself well to small, shaded rock gardens and plantings around pools and fishponds.

Spathicarpa, from *spathe,* the pale green, sheathing flower bract, and *karpos,* meaning fruit, refers to the arrangement of flowers and fruits along the narrow spadices; *sagittifolium,* from *sagitta,* meaning arrow, and *folium,* meaning leaf, refers to the long, narrow, arrowhead-like leaves.

HABIT A small, clumping, herbaceous, evergreen plant that grows to a height of about 1 foot. Densely clustered, narrow, arrowhead-shaped, shiny leaves rise from an underground rhizome. The plant has graceful, curving, cream-colored flower clusters, 3 inches long. Tiny flowers are arranged in two rows on a spadix attached to a light green spathe. The flower clusters seem to float above the leaf mass. The plant is in flower the year around except for a short period in early winter. Moderate growth rate; easily transplanted.

GROWING CONDITIONS Grows best in protected situations, in rich, well-watered, well-drained soil. Flowers best in filtered sunlight. As with most aroids, spathicarpa should not be exposed to extreme sun or beach conditions.

USE Specimen plant; mass planting; container plant.

PROPAGATION Generally propagated by root division.

INSECTS/DISEASES To control thrips, apply weak solutions of diazinon or malathion. For spider mites, spray with weak solutions of wettable sulfur. For mealybug control, use weak solutions of diazinon.

PRUNING Remove dead and damaged leaves and flowers.

FERTILIZING Apply time-release 14-14-14 fertilizer pellets to the planting bed at 4-month intervals.

DISADVANTAGES Raw plant parts may be toxic.

Spathiphyllum kochii
Spathiphyllum

The popular *Spathiphyllum kochii* is one of about 35 species of the genus that are native to Central and tropical South America. This species, primarily an ornamental landscaping plant, is the parent of two other well-known ornamental hybrids, *S. clevelandii* and the "McCoy spathiphyllum," the latter developed in Hawai'i *(S. kochii × S. cochlearispathum).*

Young flower spikes of *S. kochii* are eaten as a vegetable in tropical America, and young leaves are consumed in the Celebes. A close relative, *S. cannaefolium,* very similar in appearance, has a fragrance not unlike that of lechoso *(Stemmadenia galeottiana),* an extremely aromatic Central American tree often planted in Honolulu gardens. In tropical America *Spathiphyllum cannaefolium* leaves are dried and added to certain tobacco mixtures for aroma.

S. kochii produces lush, dark green, tropical foliage when grown in filtered sunlight. Bright, clear white flowers hover gracefully above the foliar mass in curving shapes suggesting dancers' hands. In Hawaiian gardens these plants are almost always seen in large masses. They are also excellent container specimens for shaded lanais, where other plants grow only with difficulty.

Spathiphyllum, from *spathe,* meaning the sheath, and *phyllon,* meaning leaf, refers to the white, leaflike sheath that surrounds the flower-bearing, taillike spadix; *kochii* commemorates C. Koch, a nineteenth-century botanist.

HABIT
: A herbaceous, clumping, evergreen plant that grows to about 2 feet in height; leaves grow from underground stems in dense clusters, completely covering the ground area in which they are planted. Rich, dark green leaves are long, curving, and spear-shaped. Crisp, white-spathed flower spikes, 5 inches long, are lifted above the leaf mass; seeds sometimes form along spadices; several blooms may appear concurrently. Moderate growth rate; easily transplanted.

GROWING CONDITIONS
: Extremely sensitive to strong sunlight; leaves and flowers burn unless given considerable shade. Plants are easily damaged by sea air or salty soil. Require well-drained soil; grow best in soil with a high humus content.

USE
: Specimen plant; mass planting; container plant; colorful flower spikes.

PROPAGATION
: Almost always propagated by root division; occasionally from seeds.

INSECTS/DISEASES
: For thrips, use weak solutions of malathion.

PRUNING
: Remove dead and damaged leaves and flower spikes.

FERTILIZING
: Apply general garden fertilizer (10-30-10) to the planting bed at 4-month intervals, and to container plants at monthly intervals; do not overfertilize because these tender plants will burn easily.

DISADVANTAGES
: Raw plant parts may be toxic.

Spathiphyllum kochii × *S. cochlearispathum*
McCoy Hybrid Spathiphyllum

This spathiphyllum hybrid was developed in Honolulu in 1942 by Waichi Takahashi, an orchid hybridizer for Mr. and Mrs. Lester A. McCoy, orchid fanciers. The hybrid's parents are **Spathiphyllum kochii,** and *S. cochlearis-pathum,* a native of Mexico that bears large green spathes. Of the parents, *S. kochii* is seen in great abundance in Hawaiian gardens, but the larger *S. cochlearispathum* is rare. The hybrid takes much of its size and character from the larger parent; its leaves, flowers, and height are double or more that of *S. kochii.* The green-spathed flower spikes also repeat a distinguishing characteristic of the Mexican species. This very ornamental McCoy spathiphyllum is seen in many of Hawai'i's protected gardens.

The plant is most attractively used when planted in great masses for groundcover or borders. Its huge 3-foot leaves and 8-inch flower spathes allow the plant to stand out dramatically against adjacent plants.

Spathiphyllum, from *spathe,* and *phyllon,* meaning leaf, refers to the leaflike spathe that surrounds the flower-bearing spadices of the genus; *kochii* commemorates C. Koch, original describer of the species; *cochlearispathum,* from *cochlear,* meaning spoon, and *spathe,* describes the large, green, spoon-shaped flower spathes of that species.

HABIT A herbaceous, clumping, evergreen plant that can grow to nearly 5 feet in height. Long, large leaves rise from underground stems in dense clusters, completely covering the area in which the plant is growing. Leaves are a lush moderate green, somewhat lighter in color and much larger than those of **S. kochii.** Flower spathes, about 8 inches in length, are cream-colored when young and gradually become yellow-green; they rise erect above the leaf mass. Seeds sometimes form along spadices. Moderate growth rate; easily transplanted.

GROWING CONDITIONS The plant is extremely liable to damage in exposed locations and must be protected from sun, wind, and salt burn. Grows best in moist but well-drained soils high in humus content.

USE Specimen plant; mass planting; container plant; colorful flowers.

PROPAGATION Propagated by root division.

INSECTS/DISEASES For control of thrips, use weak solutions of malathion.

PRUNING Remove dead and damaged leaves and flower spikes.

FERTILIZING Apply general garden fertilizer (10-30-10) to the planting bed at 4-month intervals, and to container plants at monthly intervals; do not overfertilize.

DISADVANTAGES Raw plant parts may be toxic.

Xanthosoma lindenii
Indian Kale

Sprung from the tropical rain forests of Colombia, the *Xanthosoma*s have been cultivated as foods since the dawn of history. Columbus found the Indians cultivating several of the 45 species native to the Americas during his first and second voyages to the New World. Most of the *Xanthosoma*s bear edible tubers, rich in starch. Several of these species have been introduced to other tropical regions throughout the world, especially by the early Spanish and Portuguese voyagers. Indomalaysians have incorporated several *Xantho-soma*s into their materia medica: some believe that the leaves will relieve fevers, others that the stems make excellent poultices for treating wounds. As with other relatives of taro, parts of the *Xanthosoma*s that are improperly prepared cause itching and burning of sensitive body tissues.

Indian kale is an ornamental relative of the several edible species. It is brightly patterned with rich green and cream-white stripes clearly differentiated on the taro-shaped leaves. This plant thrives in moist and shaded ''jungle gardens,'' and stands out in vivid contrast with plantings of darker foliage.

Xanthosoma, from *xanthos,* meaning yellow, and *soma,* meaning body, refers to the color of the interior tissues of some species; *lindenii* is named for Jules J. Linden (1817–1898), a Belgian horticulturist and publisher. Latin American names for two of the edible species are yautia *(X. jacquinii)* and malanga *(X. violaceum).* The common Hawaiian 'ape is *X. robustum.*

HABIT	A herbaceous, evergreen plant that grows to about 4 feet in height; fleshy stems sprout from underground tubers, supporting foot-long, arrowhead-shaped leaves vividly marked in green and white. Insignificant cream-colored flowers appear under the leaves. This plant is grown solely for its foliage. Fast growth rate; easily transplanted.
GROWING CONDITIONS	Prefers moist protected areas, rich soil. Although shade-loving, it must be planted where some filtered sunlight is available. Best foliar growth is achieved when the plant is protected from excessive sun, wind, and falling debris.
USE	Specimen plant; mass planting; container plant; colorful, tropical foliage.
PROPAGATION	Propagated by root division: separate vigorous young tubers from older sections and transplant young tubers in new location.
INSECTS/DISEASES	For thrips and mealybugs, use diazinon or malathion.
PRUNING	Remove dead and damaged leaves and dead flower spikes.
FERTILIZING	Apply general garden fertilizer (10-30-10) to the planting bed at 3-month intervals, and to container plants at monthly intervals.
DISADVANTAGES	Raw plant parts are toxic.

Xenophya lauterbachiana
Schizocasia

This striking plant represents one of only two species in the genus *Xeno-phya,* both of which are native to parts of Indonesia, New Guinea, and the Bismarck Archipelago. Formerly, the generic name of this species was *Schizocasia,* a term no longer acceptable to taxonomists. The name still crops up in contemporary garden conversation, however, being better known than *Xenophya.*

Like the flower spikes of its near-relative **Anthurium andraeanum,** *Xeno-phya*'s foliage looks artificial, almost as if it were made of plastic. The dramatically incised leaves, rigid and waxy, give the plant an unreal appearance. Seldom set out in a garden bed, the plant is grown most often as a container specimen.

Xenophya, from *xenius,* meaning both hospitable and foreign, and *phyas,* meaning shoot or sucker, refers to the remote habitat of the genus; *lauterbachiana* is named for Carl Adolf Georg Lauterbach (1864–1937), a German-born economist who led plant-collecting expeditions to New Guinea during the period 1899–1905. Lauterbach published several papers on the plants of New Guinea. In 1889 he undertook a voyage around the world, during which he studied some of the plants and birds living in Hawai'i. The genus *Lauterbachia* of the plant family Monimiaceae and several species of tropical plants are named in his honor.

HABIT
A herbaceous, evergreen plant that grows to about 3 feet in height. Hard, fleshy, trunklike stems support clusters of several long, narrow leaves which can be nearly 2 feet long but only 3 or 4 inches wide. Their most distinctive feature is the stylized wave pattern of the leaf edges, as if they had been carefully cut by a manicurist's scissors. Grown solely for dramatic foliage, not for flowers, which are insignificant. Moderate growth rate; easily transplanted.

GROWING CONDITIONS
Prefers humid conditions of islands in the tropical western Pacific. Grows best in moist soils, rich in humus, in places protected from extensive sun and wind.

USE
Specimen plant; container plant; dramatic tropical foliage.

PROPAGATION
Easily propagated by root division.

INSECTS/DISEASES
For thrips and mealybugs, apply weak solutions of diazinon or malathion.

PRUNING
Remove dead and damaged leaves and flower spikes.

FERTILIZING
Apply general garden fertilizer (10-30-10) to the planting bed at 3-month intervals, and to container plants at monthly intervals.

DISADVANTAGES
Raw plant parts may be toxic.

Bromeliaceae
(Pineapple Family)

Except for a single species of *Pitcairnia* from Senegal, all the bromeliads —including 250 or so other species of *Pitcairnia*—are native to tropical South America and the islands of the Caribbean. The bromeliads are a large family; 44 genera and about 1,400 species are recognized. In the American hemisphere, bromeliads are native to places ranging from the Gulf coast of the United States all the way to Cape Horn. Some dwell along the salt-drenched Atlantic and Pacific coasts; others grow in alpine habitats 14,000 feet above sea level. There are groups of desert-loving bromeliads, and also masses of rain-forest types, constantly moist and brimming with retained water. Most bromeliads are epiphytes: they grow on limbs of trees, on fallen branches, on steep cliffs, and loosened boulders without the benefit of soil. These "air plants" obtain their nourishment from the air and rain, and from the organic debris cast off by other forms of life growing around them. A smaller number are terrestrial, growing as most plants do in different kinds of soils.

The pineapple is the best known and most important member of the family. The parent species, **Ananas comosus,** has been selected and hybridized by agriculturists to produce the commercially valuable forms that are grown in such great numbers today. Several bromeliads, among them the pineapple, yield fibers of varying quality, fineness, and strength that are made into thread, twine, cordage, rope, and fabrics.

The family Bromeliaceae derives its name from one of its larger genera, *Bromelia,* which in turn is named for Olaf Bromel (1629–1705), a Swedish botanist.

Aechmea fasciata
Aechmea

The aechmeas are epiphytic bromeliads that generally grow on trees and other substrates not properly called soil. About 150 Caribbean and South American species of *Aechmea* are known. *A. fasciata,* a native of Brazil, is by far the most popular and best known species in the genus. It may also be the most popular of all garden and houseplant bromeliads, being especially noted for the beauty of its foliage and flower spikes. The plant was introduced into European horticulture in 1826, and since then has become well known throughout the world. A close relative, *A. magdalenae,* widely distributed throughout Central and South America, is the source of a strong fiber that is made into light or heavy cordage.

The aechmea is best exhibited by growing it as it is found in nature. Large colonies spreading along the branch of a tree, or grouped on tree fern trunks, offer dramatic foliar and floral displays. The species develops bright green leaves banded with streaks of lighter green. An offspring, *A. fasciata* f. Silver King, has unbanded leaves of deeper blue, that appear to be heavily dusted with talcum powder.

Aechmea, from *aechme,* meaning spear point, refers to the spines jutting outward along the edges of the leaves; *fasciata,* meaning bound together, describes the banded leaf design so distinctive of the species.

HABIT A stiffly herbaceous, evergreen plant that can grow to about 2 feet in height and diameter. Leaves develop from a short, trunklike stem, overlapping each other in a formal rosette arrangement, and curve gracefully outward and downward, producing the effect of a frozen fountain. Flower spikes rise out of the central well-like enclosure formed at the leaf bases; the pink bracts of the inflorescences produce many blue and red capsular flowers. Occasional seeds may form as flowers fade.

GROWING CONDITIONS Epiphytic in nature; does not grow well in soil; best results are obtained when roots are covered with coarse crushed tree fern fiber encased in a container, or attached to the surfaces of boulders, rock walls, or tree branches. Prefers partial shade and constant moisture from rain or sprinkler. Requires complete and perfect drainage.

USE Specimen plant; mass planting; container plant; tropical foliage and colorful flowers.

PROPAGATION Remove and replant offshoots from the base of the parent plant. May also be grown from seed: remove seeds from inflorescence and lay them, uncovered, on a bed of crushed tree fern fiber; keep moderately moist.

INSECTS/DISEASES To control scale, use malathion at one-half recommended strength; for thrips and mealybugs, use diazinon or malathion at one-half recommended strength.

PRUNING Remove old leaves from plant base and dead flower spikes only. Remove older plant crown when it begins to fade.

FERTILIZING Apply mild solutions (one-quarter strength) of foliar fertilizer at 3-month intervals to both garden and container plants.

DISADVANTAGES Those wicked, sharp spines deter some gardeners.

Ananas comosus cv. 'Red Spanish'
Pineapple, Hala Kahiki

Centuries before pineapples—canned or fresh, transported by ships or jet planes—became available to gourmets the world around, Indians of South America were enjoying the luscious fruit. In fact, the plant has been cultivated for so long over so much of tropical America that botanists today cannot determine its exact native habitat. The plant was cultivated for other reasons as well: fibers from the leaves were made into cordage and cloth, and fermented fruits yielded alcoholic beverages and vinegar. In many places a coarse but handsome cloth was woven from the leaf fibers. Spanish voyagers indirectly gave the term "pineapple" to the English-speaking world; because they thought the fruit looked like a pine cone, they called it *piña*—the name now given to the stiff fabric Filipino women weave from pineapple fibers.

Ananas comosus is one of five species in the genus. The horticultural variety known as 'Red Spanish' is pictured here. A diminutive relative is the popular dwarf pineapple, *A. nanus,* a native of central Brazil, whose perfect but tiny fruits are used in flower arrangements and table decorations. *A. comosus* var. *variegatus,* with variegated leaves, is seen often as an ornamental.

Ananas is the name the Guarani Indians of Brazil give the plant; *comosus,* hairy, describes the fine fuzz on the leaves. Hawaiians call it hala kahiki, foreign hala, because its fruits resemble those of *Pandanus odoratissimus.*

HABIT
A stiffly herbaceous, evergreen plant that grows to about 4 feet in height; long and narrow, outwardly curving gray-green leaves with barbed edges grow in rosette arrangement from a short, stocky base. A lavender-blue flower spike rises from the central leaf sheaf and swells and ripens into the characteristic fruit. Fast growth rate; easily transplanted when young. Plants fruit 18–22 months after planting.

GROWING CONDITIONS
A terrestrial bromeliad, the plant requires rich soil, generally in somewhat arid locations (about 80 inches of rain per year is best), full sun, and considerable ground moisture. Will grow and fruit in the wetter Island areas, but is more susceptible to root rot and other diseases in soils without excellent drainage.

USE
Specimen; container plant; kitchen garden plant; edible fruit.

PROPAGATION
May be propagated either from side shoots developed below fruit, or from fruit crowns. Allow shoots and crowns to dry for 30 days until roots are developing; then plant shallowly in permanent location.

INSECTS/DISEASES
For control of scale and mealybugs, use diazinon at one-half recommended strength. If root nematodes severely attack the plant, destroy it, fumigate the soil, and start a new crown or shoot.

PRUNING
Remove fruits and damaged leaves.

FERTILIZING
Before planting, apply general garden fertilizer (10-30-10) to the soil and mix well; fertilize established plants with foliar fertilizer containing nitrogen, iron, and zinc at intervals of 6 to 8 weeks.

DISADVANTAGES
Fruits poorly after two or three crops. Spiky leaves are dangerous to skin and eyes.

Billbergia macrocalyx
Billbergia

Tropical America, from southern Brazil northward through Mexico, is the primary habitat of most of the 50 or so known *Billbergia* species. The billbergias are epiphytes, growing above the forest floors on trees, fallen branches, outcroppings of rock, and other soilless substrates. They are among the most graceful and beautiful of the bromeliads; hybridization has produced many different horticultural forms. *B. macrocalyx,* now widely cultivated throughout the world, is native to Bahia State, Brazil.

The billbergias are slender, tubelike plants, little more than a sheaf of a few leaves tightly wrapped around each other, forming a narrow interior well from which the flower spikes rise. Most billbergia flower spikes become quite attenuated at maturity and have a tendency to hang gracefully over the edges of the leaf sheaf. The plants are most attractive when grown in masses on tree fern trunks or grouped along the branches or crotches of great trees. Container plants, also, are more dramatic if allowed to grow in clusters.

Billbergia is named for Gustav Billberg, a Swedish botanist; *macrocalyx,* from *macro,* meaning large, and *kalux,* cup or calyx, describes the leaflike base, normally green, that supports the petals of most flowering plants. The calyx in this species is the group of large, pink bracts along the spike.

HABIT
: A stiffly herbaceous, evergreen plant that grows to about 2 feet in height. The plant is very narrowly vertical in form; its straplike leaves are tightly bound around each other, forming an interior open tube; its leaf tips flare gracefully outward. A tall, elongated flower spike emerges from the leaf tube during the spring and summer; green and blue flowers are borne along the spike tip well above brilliant pink flower bracts. Seeds follow the bloom. Moderate growth rate; easily transplanted.

GROWING CONDITIONS
: Epiphytic in nature; does not grow well in soil, although sometimes in the wild it is found growing in soil. In cultivation, best results are obtained when the root masses are pressed into coarse tree fern fibers and set in containers or upon tree branches or other such natural sites. Prefers partial shade and constant moisture from rain or sprinkler. Roots must be placed where complete and perfect drainage is available.

USE
: Specimen plant; mass planting; container plant; colorful, tropical flowers.

PROPAGATION
: Remove and replant offshoots from base of parent plant. May also be grown from seed: remove seeds from flower spike and plant uncovered on crushed tree fern fibers and keep moderately moist.

INSECTS/DISEASES
: To control scale, use malathion at one-half recommended strength; for thrips and mealybugs, use diazinon or malathion at one-half recommended strength.

PRUNING
: Remove old leaves from plant base and dead flower spikes only. Remove older plant crown when it begins to fade.

FERTILIZING
: Apply mild solutions of foliar fertilizer at 3-month intervals to both garden and container plants.

DISADVANTAGES
: Mosquito larvae, if not controlled, may develop in the water-filled plant tubes.

Bromelia balansae
Heart of Flame

This widely cultivated bromeliad is native to Paraguay, Brazil, and Argentina. It is one of about 40 *Bromelia* species, most of which are native to areas of the Caribbean and tropical America. Like its close relative, the **pineapple (Ananas comosus)**, heart of flame is a terrestrial plant, growing in open forests and fields. Its leaves are very useful, being sources of valuable fibers. Also, because of the sharp leaf spines, they make a natural barrier when planted in a hedgerow. The fruits are eaten as such or are pressed to make juice. Two relatives, *B. pinguin* (Hawai'i's so-called wild pineapple) and *B. karatas,* also yield edible plant parts that are found in tropical American marketplaces. *B. serra* and *B. fastuosa* produce a commercial fiber, caraguata, which is woven into fabric for heavy-duty use, as in sacking and sailcloth.

Heart of flame is one of the most beautiful of the cultivated bromeliads. Its usually somewhat subdued pineapple-like central leaves turn a brilliant, fiery red at the beginning of the blooming season. Frosted white torchlike flower spikes emerge from the central well. The entire effect is captivating. In gardens the plant generally is grown as a specimen or in small masses. The lacerating leaf spines of many plants arranged in large masses make those groupings nearly impenetrable.

Bromelia is named for Olaf Bromel (1629–1705), a Swedish botanist; *balansae* honors Benjamin Balansa (1830–1891), who collected plants from several regions of the world, including Paraguay, the species' native habitat.

HABIT	A stiffly herbaceous, evergreen plant that grows to about 2 feet in height. Many narrow, jagged-edged leaves fountain out from a central stem. Normally green leaves at the plant's center turn a bright red just before the blooming season, after which a large white flower spike emerges. Individual small flowers are colored burgundy and white. Small orange edible fruits follow the blooms. Moderate growth rate; easily transplanted.
GROWING CONDITIONS	Grows best in cool, moist, partly shaded locations in soil rich in humus. Plant is rather hardy and needs no special care.
USE	Specimen plant; mass planting; container plant; colorful, tropical flowers.
PROPAGATION	Remove and replant mature offshoots from the older plant's base. Plants may also be grown from seeds.
INSECTS/DISEASES	To control scale, use malathion at one-half recommended strength; for thrips and mealybugs, use diazinon or malathion.
PRUNING	Remove dead and damaged leaves and dead flower spikes only. Remove older plant crown when it begins to fade; new offshoots will take its place.
FERTILIZING	Apply mild solutions (one-quarter strength) of foliar fertilizer at 3-month intervals to both garden and container plants.
DISADVANTAGES	Spine-edged leaves should be respected.

Cryptanthus zonatus
Earth Star, Air Plant

The genus *Cryptanthus* is native only to Brazil. Only 22 species are known, but several recognized varieties enlarge the group to nearly 50 different forms. As a group, they are very appropriately called earth stars, for most kinds of cryptanthus grow in flattened, starlike shapes quite unlike those of the majority of their bromeliad relatives. In their native habitats they can completely carpet the ground. Although they are sometimes called air plants, these plants are not epiphytes, but are terrestrial in habit, preferring humus-littered soil. They are not known to have practical uses, but are greatly admired as ornamentals.

Cryptanthus plants are small in scale, often grown in coarse tree fern fiber in small containers as table specimens or in miniature dish gardens. Several hundred plants are required in order to achieve any mass display in the garden. Singly or in masses, the form is dramatic, even somewhat other-worldly. Each species and variety is distinctive to itself in appearance. Collections of several species and varieties will produce an assortment of color and textural displays.

Cryptanthus, from *kruptos*, meaning hidden, and *anthos*, meaning flower, describes the barely visible tiny white flowers; *zonatus*, meaning banded, alludes to the striped leaves of this species.

HABIT A diminutive, leathery-leafed, evergreen plant that grows only a few inches in height and to about 6 inches in diameter, in a distinctive star-shaped form. Leaves grow in flat, open rosettes against the ground. Each leaf is clearly banded in pink and green. Tiny white flowers peek from between the leaves during the blooming season. Slow growth rate; easily transplanted.

GROWING CONDITIONS Very adaptable; will grow in moist or dry conditions provided sufficient water is made available. Grows well in full sun or partial shade; deeper and richer leaf color is achieved in somewhat shaded locations.

USE Specimen plant; mass planting; container plant; colorful foliage.

PROPAGATION Generally propagated by root division.

INSECTS/DISEASES To control scale, use malathion at one-half recommended strength. For thrips and mealybugs, use diazinon or malathion at one-half recommended strength.

PRUNING Remove dead and damaged leaves only; do not remove healthy leaves or the star-shaped rosette will be disfigured. Small flowers may be picked off as they fade. Remove older plant crown when it begins to fade; new offshoots will develop.

FERTILIZING Apply mild solutions (one-quarter strength) of foliar fertilizer at 3-month intervals to both garden and container plants.

DISADVANTAGES None in Hawai‘i.

Neoregelia carolinae
Neoregelia

The neoregelias are natives of tropical South America, chiefly of Brazil, Colombia, and Peru. Forty species comprise the genus; most of these are epiphytes, growing on trees, fallen logs, and compost-filled rock crevices. They are plants of the wet tropic jungles; their entire configuration has evolved to catch and store water in the wells formed at the bases of their leaves. Each water well is itself a microcosm, sheltering thousands of smaller plants, insects, microbes, and other kinds of minute organisms.

The Brazilian native *Neoregelia carolinae* generally is planted singly in a container holding tree fern fiber or rich soil, or else on decomposing stumps or logs in an informal jungle arrangement. Individual plants should be allowed plenty of growing space, so that the perfect fountain of leaves each produces may develop without restriction. As with plants of several other bromeliad genera, neoregelia foliage exhibits a variety of brilliant hues, thereby presenting an exciting background for the flower clusters that soon follow. Neoregelia flowers develop and bloom deep within the water wells.

Neoregelia derives from *neo,* meaning new, and *regelia,* an older genus named for Dr. Eduard A. von Regel (1815–1892), a botanist and superintendent of the Imperial Botanical Gardens, at St. Petersburg, Russia; *carolinae* is named for M. H. Carolyn V. Morren, wife of one of the nineteenth-century editors of *La Belgique Horticole,* in Liège, Belgium.

HABIT
: A supple, yet leathery, herbaceous plant that grows to about 1 foot in height and 18 inches in diameter. Produces a rosette of tightly overlapping leaves that form a well at the plant's center. Foot-long leaves bend gracefully outward and then inward. Before the blooming period, the foliage, normally striped in pink and green, turns a brilliant red; following this color change, clusters of small lavender flowers appear deep within the water-filled well. Slow growth rate; easily transplanted.

GROWING CONDITIONS
: Requires a great deal of moisture, either from natural rainfall or from overhead irrigation. Best foliage is produced in partial shade and protected locations.

USE
: Specimen plant; mass planting; container plant; colorful and dramatic tropical foliage and flowers.

PROPAGATION
: Generally propagated by root division; remove offshoots when mature and transplant to new locations. Keep constantly moist.

INSECTS/DISEASES
: To control scale, use malathion at one-half recommended strength. For thrips and mealybugs, use diazinon or malathion at one-half recommended strength. Flush out water wells 2 or 3 times weekly to control mosquito larvae.

PRUNING
: Remove dead and damaged leaves and old crowns.

FERTILIZING
: Apply mild solutions (one-quarter strength) of foliar fertilizer at 3-month intervals to both garden and container plants.

DISADVANTAGES
: Water wells may harbor mosquito larvae.

Neoregelia melanodonta
Black-toothed Neoregelia

Because of its exotic colors, *Neoregelia melanodonta* is one of the plants most widely sought for Hawai'i's bromeliad gardens. A native of Brazil's forests near Espírito Santo, the plant can now be found in tropical gardens across the world. This neoregelia's epiphytic habit is similar to that of most species in the genus; in its native home it grows on trunks of trees, on fallen logs, or on humus-littered outcroppings of rock.

People in tropical America buy these plants in the marketplace or go to the forests to collect them for household decorations. This is done especially at Christmas and Easter time, when special floral decorations are arranged. Often the plants will be thrown upon the tile roofs of the houses, where they attach themselves readily, creating colorful hanging gardens throughout the villages.

Neoregelia derives from *neo,* meaning new, and *regelia,* an older genus named for Dr. Eduard A. von Regel (1815–1892), a botanist and superintendent of the Imperial Botanical Gardens, at St. Petersburg, Russia; *melanodonta,* from *melanos,* meaning black, and *dens,* meaning tooth, describes the black-toothed edges of the leaves.

HABIT A supple, yet leathery, herbaceous plant that grows to about 1 foot in height and 18 inches in diameter. Produces a rosette of tightly overlapping leaves that form a well at the plant's center. Tooth-edged, foot-long leaves bend gracefully outward and backward. Before the blooming period, the leaves, normally bright green, turn to brilliant shades of red and red violet. Bright blue flowers appear in a dense cluster deep within the water-filled well. Slow growth rate; easily transplanted.

GROWING CONDITIONS An epiphyte by preference, the plant must be grown on branches of trees or in substrates rich in humus, such as coarse crushed tree fern fiber and cinders. Requires a great deal of moisture, either from natural rainfall or from overhead irrigation. Best foliage is produced in protected locations and partial shade.

USE Specimen plant; mass planting; container plant; colorful and dramatic tropical foliage and flowers.

PROPAGATION By root division.

INSECTS/DISEASES To control scale, use malathion at one-half recommended strength. For thrips and mealybugs, use diazinon or malathion at one-half recommended strength. Flush out water wells 2 or 3 times weekly to control mosquito larvae.

PRUNING Remove dead and damaged leaves and old crowns.

FERTILIZING Apply mild solutions (one-quarter strength) of foliar fertilizer at 3-month intervals to both garden and container plants.

DISADVANTAGES Water wells may harbor mosquito larvae.

Tillandsia cyanea
Tillandsia

Tillandsia cyanea is one of 500 species of the genus that are native to the warm temperate and tropic regions of the Americas. Many tillandsias are gathered for sale in Central and South American markets, especially at Christmas and Easter time. Several of the tillandsias are known in Central America as pie de gallo. The term means rooster foot and describes the clawlike flowers of those species. Often the tillandsias are planted in gardens and in pots for courtyard color. In many Central American villages there are rooftop gardens of bromeliads in which the tillandsias are included; the owners simply throw the plants upon the tiled roofs, where these ''air plants'' take root and grow prolifically.

In Hawai'i, *T. cyanea,* an epiphytic native of Ecuador and Guatemala, generally is planted on tree fern trunks or in pots filled with tree fern fiber. It is one of the most popular of bromeliads for landscaping because of its interesting foliage and beautiful flower spikes, and because, unlike certain other bromeliads, its crown does not hold water in which mosquitoes can breed.

Tillandsia is named for Elias Tillands (1640–1693), a Swedish physician, botanist, and professor of medicine at the University of Åbo, Finland, who listed many of Finland's plants in a botanical work published in 1673; *cyanea,* meaning dark blue, refers to the color of the flowers.

HABIT — A stiffly, herbaceous evergreen plant that grows to about 18 inches in both height and diameter. Many long, thin gray-green leaves hang from the central stem. During the blooming period in winter and spring, brilliant pink spikes of close-packed bracts appear; the blue-violet flowers develop soon after. Slow growth rate; easily transplanted.

GROWING CONDITIONS — Naturally an epiphyte; must be grown in coarse compost, such as decomposing tree fern fiber. Prefers areas with considerable moisture; roots must have excellent drainage. Blooms best in filtered sunlight. Grows very well at Hawai'i's higher elevations.

USE — Specimen plant; mass planting; container plant; colorful flower spikes.

PROPAGATION — Remove and replant mature offshoots from the plant's base.

INSECTS/DISEASES — To control thrips, apply malathion at one-half recommended strength.

PRUNING — Remove dead and damaged foliage and dead flower spikes only; do not crop back healthy foliage because cut leaves will not be replaced. Remove the old crown entirely when it begins to fade, so that new offshoots may take its place.

FERTILIZING — Apply mild solutions (one-quarter strength) of foliar fertilizer at 3-month intervals to both garden and container plants.

DISADVANTAGES — None in Hawai'i.

Tillandsia usneoides
Spanish Moss, Dole's Beard, Hinahina

Spanish moss is not a moss at all, but a very unusual bromeliad. It is one of the most widely dispersed of the pineapple's relatives, growing in great abundance from the Gulf region of the southern United States throughout much of Central and South America to the cooler regions of Argentina and Chile. It is very characteristic of southern forests in the United States, where it hangs in gray-green profusion from the boughs of host plants, such as pines, cypresses, and live oaks.

Spanish moss has long been used as a stuffing and packing material. Sometimes it is called false horsehair or vegetable horsehair, because it makes a satisfactory substitute for that kind of stuffing. This light, dry, filamentous material has even been used in filters for air conditioners and for oil.

Hawaiians call the plant hinahina, which means gray, and plait it into beautiful open-ended leis. Another local name for the plant is Dole's beard, a reference to the hirsute adornment worn by Sanford Ballard Dole (1844–1926), one of the founders of the Republic of Hawai'i, and, after Annexation, first governor of the Territory.

Tillandsia is named for Elias Tillands (1640–1693), a Swedish physician, botanist, and professor of medicine at the University of Åbo, Finland; *usneoides,* meaning like *Usnea,* refers to the fact that this bromeliad has leaves similar in appearance to parts of the lichen genus *Usnea.*

HABIT
A diminutive, mosslike, herbaceous, "evergreen" plant that grows in colonies of varying mass; individual sections can be quite small. Purely epiphytic in habit, hanging pendent from natural or man-made supports, the plant takes no nourishment from its host but lives off decomposed organic residue that gathers on the surface of the host. Soft, small gray leaves; very tiny, insignificant chartreuse flowers. Moderate growth rate; easily transplanted.

GROWING CONDITIONS
Very adaptable; will grow almost anywhere, except in the teeth of strong salt winds. Grows well in either moist or dry areas with adequate and fairly regular supply of moisture. Will survive considerable neglect. Best growth is obtained in partly shaded areas protected from winds.

USE
Specimen plant; mass planting; hanging basket plant; unusual gray foliage.

PROPAGATION
Almost any piece of the plant will grow if it is firmly attached to some support.

INSECTS/DISEASES
None of much importance in Hawai'i; occasionally turns black from fungus disease if kept too wet. Remove affected parts and spray remaining sections with mild solutions of captan.

PRUNING
Remove excess parts of plant only as needed to limit size or to propagate.

FERTILIZING
Apply weak solutions (one-quarter strength) of foliar fertilizer at 3-month intervals to entire plant mass.

DISADVANTAGES
None in Hawai'i.

Vriesea hieroglyphica
Vriesea, King of the Bromeliads

Vriesea hieroglyphica, a native of Brazil, sometimes is called the king of the bromeliads because of its grand size. It is one of about 190 *Vriesea* species found throughout tropical America. Vrieseas are plants of the tropical rain forest, growing best where high humidity, rainfall, or clouds are continually present. Some species grow near sea level, while others inhabit ridges and mountainsides up to altitudes of 10,000 feet. The vrieseas have no value except that of their great beauty as ornaments to homes and gardens.

Vrieseas are epiphytic, growing in the crotches of rain-forest trees or perching on decaying stumps or logs. In Hawai'i the plants generally are attached to tree fern logs, or are planted in pots filled with coarse but densely packed tree fern fiber. *V. hieroglyphica* grows to such a great size that generally it is shown as a specimen. Its kingly presence is commanding, indeed; seldom is it overlooked by even the most casual eye.

Vriesea is named for W. H. de Vries (1806–1862), a Dutch botanist; *hieroglyphica,* meaning like a hieroglyph, describes the foliage, distinctively marked and banded as if it were inscribed with an ancient kind of writing.

HABIT	A supple, yet leathery, herbaceous evergreen plant that grows to about 5 feet in height and 6 feet in diameter. Produces a large rosette of tightly overlapping leaves, which form a good-sized water well at their center. Smooth and untoothed leaves bend gracefully outward and backward. The long, strap-shaped leaves retain a handsome combination of brilliant greens banded with purple throughout the plant's life. The species has seldom blossomed in Hawai'i, but in other areas, tall, soft, green and yellow flower spikes often grow from the water well. Slow growth rate; easily transplanted.
GROWING CONDITIONS	Being an epiphyte, the plant must be grown in planting mixtures rich in humus, such as coarse crushed tree fern fiber with cinders. Requires a great deal of moisture, either from natural rainfall or from overhead irrigation. Best foliage is produced and maintained in protected locations with partial shade. Grows very well if attached to tree crotches on a small bed of tree fern fiber.
USE	Specimen plant; container plant; dramatic tropical foliage; seldom grown in masses because of its size.
PROPAGATION	From root offshoots.
INSECTS/DISEASES	To control scale, use malathion at one-half recommended strength. For thrips and mealybugs, use diazinon or malathion at one-half recommended strength. Flush out water wells 2 or 3 times weekly to control mosquito larvae.
PRUNING	Remove dead leaves—but with care to avoid damaging others.
FERTILIZING	Apply weak solutions (one-quarter strength) of foliar fertilizer at 3-month intervals to both garden and container plants.
DISADVANTAGES	Water wells may harbor mosquito larvae.

80

Commelinaceae
(Spiderwort Family)

The spiderworts in general are a tropical and subtropical family; representatives grow naturally in many of the world's warmer regions. A few will grow in warmer parts of the temperate zones, but most are killed by frost; tropical species grown in northern latitudes must be protected in greenhouses or interiors of residences during winter months. One North American native, *Tradescantia virginica,* is used as an herb for cooking and in salads. An Asian relative, *Commelina benghalensis,* is eaten as an emergency food in times of famine. This plant and a near relative, *C. nudiflora,* have mucilaginous leaves which are used for poulticing sores in Malaysia. Several species are gathered for animal fodder in Southeast Asia. Best known of Hawai'i's ornamental spiderworts are the wandering Jews (*Tradescantia* spp.), oyster plant *(Rhoeo spathacea),* **blue ginger *(Dichorisandra thyrsiflora),*** and honohono ''grass'' *(Commelina diffusa).*

The family is based on the genus *Commelina,* named in honor of two Dutch brothers, Kaspar Commelin (1667–1731) and Johann Commelin (1629–1698), both of whom were botanists.

Campelia zanonia
Campelia

This spiderwort is one of three species of *Campelia* that are native to the Caribbean and tropical America. All are said to bear edible fruits. Like most of its spiderwort relatives, this campelia is rather tender and herbaceous. Sometimes it is mistaken for certain of the **dracaenas** of the **lily family (Liliaceae)** because of similar long, striated leaves. As with all spiderworts, however, campelia flowers are distinctively tricornered.

C. zanonia is decidedly a tropical plant, succulent in appearance and rather unrestrained in growth. Its colorful pink, green, and cream leaves bring bright spots of color into shady garden nooks, where it can grow quite happily. Campelia is probably best planted in masses as a border plant or for low background groupings. It mingles well with other tropical plants, such as gingers, heliconias, and ferns.

Campelia, from *kampe,* meaning a binding, refers to the clasping leaf bases; *zanonia* is named for Giacoma Zanoni (1615–1682), prefect of the botanic garden at Bologna, and author of *Iostoria Botanica,* published in 1675.

HABIT	A loose-structured, succulent, evergreen plant that grows to about 3 feet in both height and diameter. Leaf stalks are composed of overlapping leaf sheaths; a new leaf sprouts from within the fold of the one preceding it. Individual leaves are rather long, narrow, and slightly limp, giving the impression of a tender frailty. Foliage is prettily colorful, mostly a creamy yellow with variegations of green stripes and pink edging. Small white tricorn flowers appear constantly in small clusters. Tiny seeds follow the bloom. Fast growth rate; easily transplanted when young; rather difficult to transplant without injury when plants are large and established.
GROWING CONDITIONS	Shade-loving and rather reclusive in nature, the plant enjoys similar conditions in cultivation. Grows best in partial shade, in soils rich in humus material. Requires large amounts of water, but must be allowed good drainage. Should be planted in places where heavy traffic or abrasion will not injure the foliage.
USE	Specimen plant; mass planting; colorful tropical foliage.
PROPAGATION	Easily propagated by cuttings.
INSECTS/DISEASES	To control thrips, apply diazinon or malathion at one-half recommended strength.
PRUNING	In general, remove only dead and damaged plant parts. If leaf stalk is badly damaged, unsightly, or too long, cut entire stalk back to ground and allow new shoot to take its place.
FERTILIZING	Apply general garden fertilizer (10-30-10) to the planting bed at 3-month intervals.
DISADVANTAGES	None of any consequence.

Dichorisandra thyrsiflora
Blue Ginger

This is one of those plants that looks like something it is not. Although very similar to many of the gingers of the family **Zingiberaceae,** actually it is simply a very large member of the spiderwort family. In truth, this import from Brazil should be called blue spiderwort instead of blue ginger, but in ginger-conscious Hawai'i, people insist on including it within that spectacular group.

This "blue ginger," then, is one of thirty-five species of *Dichorisandra,* all of which are native to tropical America. Two common ornamental relatives are *D. reginae,* native to Peru, grown for its colorful foliage, and *D. hexandra,* a flowering tropical herb growing throughout much of tropical America. Both these relatives can be found in some of our Hawaiian gardens.

Blue ginger generally is treated in much the same way as are the true gingers. In Hawai'i's gardens it grows to heights well above that of a man's head, and its brilliant flowers are among the bluest of all we see here.

Dichorisandra, from *di-,* meaning twice or double, *chorizo,* meaning to part, and *aner,* meaning man, describes the two-valved structure of the anthers; *thyrsiflora,* from *thyrsus,* is a botanical term that describes a flower cluster which is larger in the center than at the top and bottom (somewhat like a football in profile).

HABIT
A succulent, herbaceous, evergreen plant that grows to about 8 feet in height. It can cover rather large areas in much the same way as can the gingers. Leaf stalks about 1 to 2 inches thick emerge from the underground stems and grow vertically, with leaves and flower clusters at the tops. Leaves are quite large, to about 8 inches in length, and are green with silvery striations. Flower panicles appear periodically with many florets to each panicle; individual tricornered blue florets are about one-half inch in diameter. Moderate growth rate; easily transplanted when small.

GROWING CONDITIONS
Prefers cool, moist, shaded, protected areas where soil is rich in humus. In Hawai'i grows best in inner valleys. In general not a good beach plant, but with sufficient protection from sun and sea wind it will produce respectable foliage and flowers.

USE
Specimen plant; mass planting; colorful tropical flowers.

PROPAGATION
By root division or stem cuttings.

INSECTS/DISEASES
To control thrips, apply diazinon or malathion at one-half recommended strength.

PRUNING
Remove dead and damaged leaves and flowers; unsightly stalks may be cut completely to the ground.

FERTILIZING
Apply general garden fertilizer (10-30-10) to the planting bed at 3-month intervals.

DISADVANTAGES
May invade garden areas where it is not wanted.

Liliaceae
(Lily Family)

The lily family is one of the largest in the plant kingdom, containing about 250 genera and more than 3,700 species, which grow in most of the world's climatic regions. Formerly, taxonomists included in the family such diverse plants as onions, yuccas, **dracaenas, sansevierias, agaves,** alstroemerias, **cordylines,** and trilliums—all of them now assigned to several other families. Most common of the ornamental lilies are members of the genera *Gloriosa, Colchicum, Funkia, Hemerocallis, Aloë, Gasteria, Haworthia, Lilium, Tulipa, Fritillaria, Scilla, Hyacinthus, Astelia, Asparagus, Convallaria,* and *Ophiopogon.* Of these, day lilies, hyacinths, tulips, crocuses, gloriosa lilies, Easter lily, and asparagus must be the best known. Mondo grass *(Ophiopogon japonicus)* is the lily most often planted in Hawai'i.

Lilium, the generic basis for the family's name, is the Latin equivalent of the Greek *leirion,* the name that Theophrastus (third century B.C.), the "father of botany," gave to the madonna lily, *(L. candidum).*

Asparagus africanus
Asparagus, Regal Fern

Regal fern is one of those incorrect names that really should not be continued. Neither this plant nor any of the other asparaguses belongs to the fern group; yet people very often refer to them as "asparagus ferns." A much more correct appellation is, simply, asparagus.

Asparagus africanus is a native of South Africa. It belongs to a genus of about 300 species, all originating in the Old World. Women of several South African tribes use the leaves of this species in preparing a pomade to encourage growth of hair. The most valuable species is *A. officinalis,* a native to the Old World's temperate zone, which produces the familiar vegetable sold in the world's markets. Several relatives have similar, but more local, market value; among them are *A. acutifolius* from the Mediterranean region; *A. albus,* known in French-speaking Africa as "asparagus sauvage"; and *A. cochinchinensis* from China and Japan.

A. africanus is an important plant for florists, being used almost universally as a filler and background material in commercial flower arrangements. Although it is well behaved as a potted specimen, in the garden this plant takes on a much more irregular and unencumbered character.

Asparagus is an old Greek name used by Theophrastus, the "father of botany"; *africanus* indicates the native habitat of this species.

HABIT An upright, somewhat woody, evergreen plant that grows to about 6 feet in height. Several stems rise from underground rhizomes and branch out to form thickly foliated tops; individual leaves are very small, appearing along the stems in tight whorls in much the same way as do those of some conifers. Tiny white flowers appear along the branches periodically, followed by bright red berries. Fast growth rate; easily transplanted.

GROWING CONDITIONS Very adaptable; will grow in almost any location except where exposed to salt winds. Enjoys rich, well-watered, well-drained soils in areas of full sun, but will tolerate a considerable amount of shade. A very hardy plant.

USE Specimen plant; container plant; long-lasting cut foliage.

PROPAGATION May be grown from seeds or can be propagated easily by root division.

INSECTS/DISEASES For control of mealybugs on foliage, use diazinon or malathion. For root mealybug control, drench planting soil with malathion solution. Severely infested plants should be discarded and the immediately surrounding soil should be fumigated or heavily drenched with malathion to eradicate the mealybug colony.

PRUNING May be severely pruned to shape the plant or to use branches as arrangement materials.

FERTILIZING Apply general garden fertilizer (10-30-10) to the planting bed at 3-month intervals, and to container plants at monthly intervals.

DISADVANTAGES Hidden thorns sometimes snare the unwary.

Asparagus densiflorus cv. 'Myers'
Myers' Asparagus

This plant is known generally in the nursery trade as *Asparagus myersii.* The term has no botanical standing, because the plant is simply a horticultural cultivar of the South African species, *A. densiflorus.* The species seems to be purely ornamental; however, many others of the 300 or so *Asparagus* species are used as foods or as medicines. Seeds of the popular vegetable, *A. officinalis,* are used sometimes as a coffee substitute. The roots of an Indian and Australian relative, *A. racemosus,* are said to be useful in the treatment of dysentery and urinary problems; also preparations from the roots are applied as ointments for soothing inflamed tissues.

Myers' asparagus is quite unlike Hawai'i's other more open-formed asparaguses in that its stems are very compact, resembling elongated cones that are densely foliated. These conical masses emerge from the roots in a disciplined, rayed arrangement. It is a gardener's joy, always attractive, almost self-sufficient, requiring only average amounts of water and thriving in one location for many years.

Asparagus is a name used by Theophrastus, the "father of botany"; *densiflorus,* from *densus,* meaning close or crowded, and *flos,* meaning flower, describes the myriad tiny starlike blossoms that crowd the leaf stems.

HABIT A low, clumping, herbaceous, evergreen plant that grows to about 3 feet in height; older plantings spread out over a considerable area. Conical stems with dense, fine foliage emerge from the root mass in a sunray formation. The entire plant is very compact and dense. Periodically, the plant bursts into bloom; tiny white flowers appear among the even tinier leaf clusters; occasional red berries follow. Moderate growth rate, easily transplanted.

GROWING CONDITIONS Quite adaptable and hardy; will withstand a certain amount of inattention, but grows best with regular watering. Prefers rich, well-watered, well-drained soil, full sun or some shade. Plants in sunny locations present a much more dense and compact habit; specimens growing in shade become somewhat attenuated.

USE Specimen plant; mass planting; container plant; dramatic foliage.

PROPAGATION May be grown from seeds or may be propagated easily by root division.

INSECTS/DISEASES For control of mealybugs on foliage, use diazinon or malathion. For root mealybug control, drench planting soil with malathion solution. Severely infested plants should be discarded and the immediately surrounding soil should be fumigated or heavily drenched with malathion to eradicate the mealybug colony.

PRUNING Requires little pruning except when an entire stem dies.

FERTILIZING Apply general garden fertilizer (10-30-10) to the planting bed at 3-month intervals, and to container plants at monthly intervals.

DISADVANTAGES None.

Aspidistra elatior var. *variegata*
Aspidistra

Aspidistra probably is one of the world's most often used and least cared-for house plants. Nongardeners learned long ago that this hardy greenery, once settled on a parlor table near a window, would live there for years with only the most casual attention. Surely, aspidistra must rank along with **bowstring hemp *(Sansevieria trifasciata)*** and the parlor palm *(Collinia elegans),* as a champion among long-lived house plants. This particular aspidistra is a native of China; seven other species of the genus are found in parts of Eastern Asia. Those species cultivated by man seem to be strictly ornamental.

The habit of growing aspidistra as a potted plant is so ingrained among Westerners that almost never do they think of showing it in any other way. Actually, it makes an attractive groundcover of moderate height for tropical gardens. Its bold and handsome leaves provide a rich contrast for ferns and other fine-leafed tropicals. It is a shade lover and grows very well in darker, sun-robbed areas of gardens—as it does in parlors.

Aspidistra, from *aspidiseon,* a small round shield, probably refers to the shape of the flower's stigma; *elatior* means taller (than other aspidistras); *variegata* alludes to this variety's colorful foliage.

HABIT
A clumping, herbaceous, evergreen plant that grows to about 18 inches in height. Dark green and somewhat crinkled, paddle-shaped leaves are 18 to 20 inches long; this variety has leaves with white striations. Small, brownish flowers hide among the foliage. Moderate growth rate; easily transplanted.

GROWING CONDITIONS
Very adaptable; will grow almost anywhere except in strong salt winds. Must be protected somewhat from too much direct sunlight because foliage is likely to discolor or burn. Prefers rich, well-watered, well-drained soils; however, it will manage to exist even with very poor soil and minimum care.

USE
Specimen plant; mass planting; container plant; large, attractive foliage.

PROPAGATION
Propagated easily by root division.

INSECTS/DISEASES
To control scale, use malathion at one-half recommended strength. For thrips and mealybugs, use diazinon or malathion at one-half recommended strength.

PRUNING
Prune dead and damaged leaves only; plant is extremely neat and tidy, requires little attention. Entire plant may be rejuvenated by cutting all leaves and trunk back to ground level.

FERTILIZING
Apply general garden fertilizer to the planting bed at 4-month intervals, and to container plants at monthly intervals.

DISADVANTAGES
None.

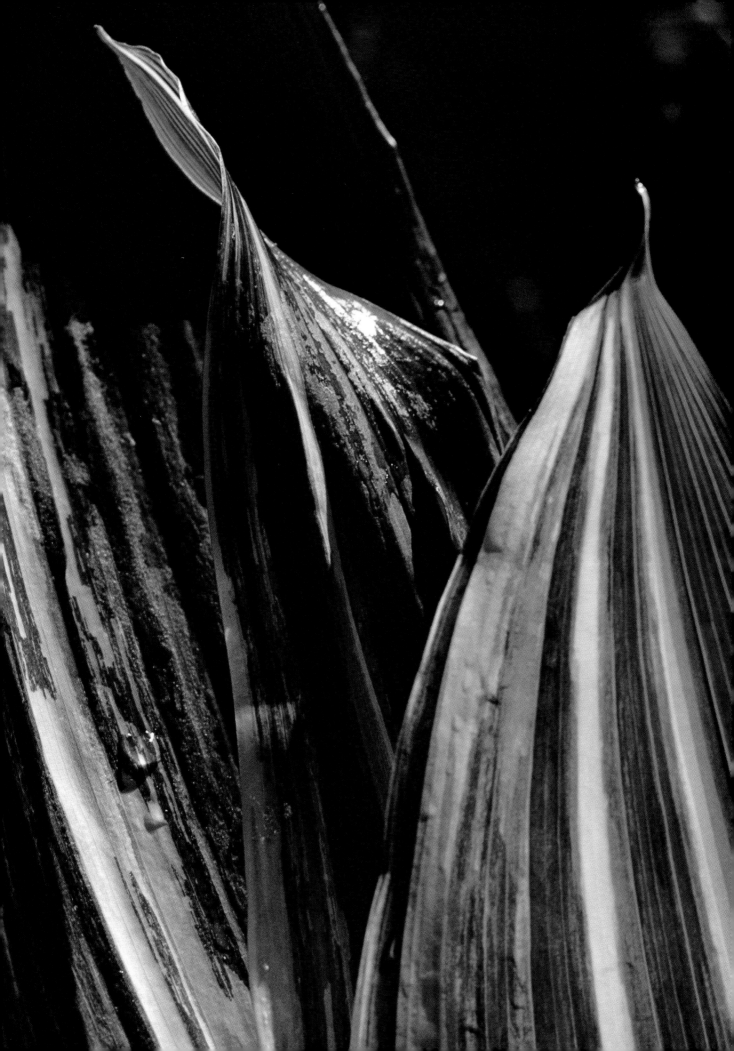

Agavaceae
(Agave Family)

At one time, plants in the agave family were considered to be members of the **lily family.** Today, the family Agavaceae consists of 20 genera and about 670 species, most of which are either tropical or subtropical in origin. Chief among the genera are *Agave, Cordyline, Dracaena, Furcraea, Nolina, Phormium, Polianthes,* and *Yucca.* Many of the plants in these genera are useful to mankind. Maoris have used New Zealand flax, *Phormium tenax,* as the source of their chief fiber since first they settled these islands. Hawaiians—and all their Polynesian cousins—have cultivated **ti *(Cordyline terminalis)*** for use in packaging foods and for shelter, clothing, and sweetmeats, and (after Westerners taught them how) for making an alcoholic beverage. Mexicans developed the beverages pulque and mescal from the maguey plant, *Agave americana.* A close relative from Central America, *A. sisalana,* is the source of sisal, one of the world's most important heavy-duty fibers, from which strong ropes and hawsers are woven.

The term Agavaceae is derived from the name of the genus *Agave,* which comes from the Greek word *agavos,* meaning admirable, probably referring to the handsome flower spikes produced by most plants in the group, but also referring perhaps to the many practical uses to which these plants can be put.

Cordyline terminalis
Red Ti

Red ti is native to Malaysia, Indonesia, and other tropical South Pacific islands. It is almost identical with green ti (also classified as *Cordyline terminalis*), except for the notable fact that it has kept the ability to produce seeds, whereas the green ti has lost this ability. Because of this important difference, the red ti is able to produce progeny showing many different and beautiful leaf variations which enthusiasts collect with much pleasure and sometimes at considerable expense. Many Hawaiians believe that a red ti grown near the home is an open invitation to misfortune. Malays believe the opposite: to them, a red-leafed ti dispels evil spirits.

Red ti's foliar forms are legion. The leaves may be wide or thin, short or long, small or large, single colored or many hued. The names for popular forms are descriptive and imaginative—Pele's Smoke, Hawaiian Flag, Hilo Rainbow, Kahana, Menehune, Mauna Kea Snow, Maui Beauty. The foliage, favored for use in flower arrangements, is shipped in ''tropical bouquets'' from Hawai'i to many parts of of the world. The variety called Onomea is one of the leading foliage plants in Hawai'i's floral trade.

Cordyline, from *kordyle,* meaning a club, describes the large, fleshy, clublike roots; *terminalis,* meaning at the end, refers to the fact that the flower clusters are borne at the tops of the leaf crowns. The Polynesian word ti is the accepted general term used throughout the Pacific. The Hawaiian variant, written *ki,* is more properly used in these islands.

HABIT
: An erect, woody, evergreen plant that can grow to 12 feet or more in height. Single or multiple stems support clusters of 18-inch-long, glossy, spatulate leaves. Periodically, clusters of large fragrant pink or light red-violet flowers emerge at the tops, followed by occasional bright red berries. Moderate growth rate; easily transplanted.

GROWING CONDITIONS
: Very adaptable; will grow in almost any place that is protected from heavy winds. Grows best in cool, moist, wind-free locations, in soil rich in humus and with constant water supply. The red-leafed forms are best grown in partial shade.

USE
: Specimen plant; mass planting; container plant; colorful tropical foliage.

PROPAGATION
: Seeds may produce new varieties; established foliage forms are propagated by cuttings.

INSECTS/DISEASES
: To control scale, apply summer oil or malathion. For mealybugs, use diazinon or malathion. For Chinese rose beetles, spray with one of the residual insecticides. During very wet periods leaves may develop numerous yellow spots due to a fungus disease; to prevent spread; remove affected leaves. Control by fungicides is difficult.

PRUNING
: Remove faded leaves; plant may be severely pruned to rejuvenate it.

FERTILIZING
: Apply general garden fertilizer (10-30-10) to the planting bed at 3-month intervals, and to container plants at monthly intervals.

DISADVANTAGES
: Leaves are easily shredded by the wind, falling debris, and Chinese rose beetles.

Dracaena concinna
Money Tree

This native of Madagascar is undoubtedly the most popular of all dracaenas grown in Hawai'i. It even outranks Hawai'i's much respected native hala-pepe *(Dracaena aurea).* A majority of Island gardens have at least one money tree planted somewhere within their environs. Hawai'i's first representatives of this species seem to have been introduced to the Big Island sometime in the early 1920s. The exciting newcomer gained its local name when some of the plants were set out in the grounds of the old Bishop Bank in Hilo. The proximity of new plant and established bank quickly led people to associate the plant with money. This tree is commonly called *D. marginata.*

This striking and dramatic plant ultimately assumes the proportions of a tree; however, most gardeners prune or otherwise restrain it to the size of a shrub. It assumes varied forms with training, and, under control, is a very suitable selection for narrow or otherwise restricted areas. It is also very hardy, and will withstand much neglect.

Dracaena, from *drakaina,* meaning female dragon, is the generic term first associated with the dragon's blood trees *(D. cinnabari* and *D. schizantha); concinna,* meaning neat and elegant, describes the species' ordered habit.

HABIT   An erect, angular, woody, evergreen plant that can grow to about 30 feet in height. Older plants are characteristically bare-stemmed and branched, inasmuch as the foliage appears at or near the growing tips. Individual leaves, about ½ inch wide and 24 inches long, are rapier-shaped. Leaves are rich green for the most part, with distinguishing red margins. Large, open, erect, triangular flower clusters of a yellow-orange color appear periodically; orange fruits containing black seeds form after the bloom. Moderate growth rate; easily transplanted.

GROWING CONDITIONS   Very adaptable; will grow almost anywhere in Hawai'i, even at the beach if protected from the strongest salt winds. Prefers rich, well-watered, well-drained soil, and full sun or partial shade.

USE   Specimen plant; mass planting; container plant; tropical foliage.

PROPAGATION   May be grown from either cuttings or seeds.

INSECTS/DISEASES   To control scale, apply summer oil or malathion. For mealybugs, use diazinon or malathion.

PRUNING   May be drastically pruned to control size and shape; new shoots eventually will appear at the cut end.

FERTILIZING   Apply general garden fertilizer (10-30-10) to the planting bed at 3-month intervals, and to container plants at monthly intervals.

DISADVANTAGES   Sometimes the plant, rather large at maturity, is set in places that are too small to hold it.

Dracaena concinna cv. 'Tricolor'
Tricolor Dracaena

Dracaena concinna is a picturesque plant from the island of Mauritius. It is one of about 150 *Dracaena* species native to the tropics of the Old World. The tricolored dracaena is a horticultural variety that has been developed fairly recently for the nursery trade. In 1969 it was brought from Japan to Hawai'i, where several nurseries now propagate it for sale.

The cultivar is quite different from its parent: the foliage is much more highly colored, with rich, creamy white variegations lighting the bright green background, and the brilliant red margins are much brighter and more pronounced than are those of the parent. In the garden the tricolor cultivar is used in much the same way as is the money tree. It is less stiff, much bushier, and a bit more graceful than is that handsome and dramatic relative.

Dracaena, from *drakaina,* meaning female dragon, refers to the close botanical relationship with the dragon trees, *D. cinnabari* and *D. schizantha; concinna,* meaning neat and elegant, describes the species' ordered habit.

HABIT An erect, woody, evergreen plant that grows to about 8 feet in height. Single or multiple stems sprout from the root mass; stems initially produce single foliage heads, but as plant develops leafy branches will form. Brilliant green and creamy white foliage is edged in red; leaves, softly curving and pliable, grow to about 8 inches in length. Small yellowish flowers appear in erect, open, foot-high clusters and are followed by small round fruits.

GROWING CONDITIONS The plant is quite adaptable but grows at its best in partly shaded places protected from wind. Requires rich, well-watered, well-drained soil. Not a beach plant.

USE Specimen plant; container plant; colorful tropical foliage.

PROPAGATION Generally propagated by stem cuttings, but may be started from seeds when available.

INSECTS/DISEASES To control scale, apply summer oil or malathion. For mealybugs, use diazinon or malathion.

PRUNING Remove dead and damaged foliage only; drastic pruning of entire leaf cluster may leave unsightly stumps; new crowns will sprout near the stump tops, but plant may be less attractive as a result.

FERTILIZING Apply general garden fertilizer (10-30-10) to the planting bed at 3-month intervals, and to container plants at monthly intervals.

DISADVANTAGES None.

Dracaena draco
Dragon Tree

The valuable resin known to early Mediterranean peoples as "dragon's blood" came from two plants: *Dracaena cinnabari,* from the island of Socotra in the Indian Ocean, and *D. schizantha,* from southern Arabia and East Africa. Somewhat later *D. draco* was discovered in the Canary Islands and became an additional source for the brilliant red resin. Dragon's blood has been highly prized in making fine varnishes over the centuries and was an important item of trade throughout the Indian Ocean and the Mediterranean Sea long before the time of Christ. People of those times thought that the resin was indeed the residue of real blood lost by dragons in great contests among themselves or with other animals. Latin and French names for the resin were *sanguis draconis* and *sang de dragon.*

Quite often in Hawai'i the dragon tree is employed dramatically in somewhat stark arrangements, often against a background of pebbles and rocks and in conjunction with other dryland plants. It is rather stiff in character, yet very beautiful, with its gracefully arrayed leaves shining silvery against the more usual garden greens. It is a popular plant for small or narrow areas that have room for only one specimen plant. Flower arrangers collect the dried leaves, generally displaying them bottoms-up, in order to reveal the brilliant orange inner leaf sheaths.

Dracaena, from *drakaina,* meaning female dragon, and *draco,* also meaning dragon, emphasize the plant's capacity to produce dragon's blood resin.

HABIT	A woody, evergreen, treelike plant that can grow to more than 30 feet in height. Generally has a large central trunk that with age may develop a number of side branches; both trunk and branches produce large, compact clusters of silvery gray leaves, each about 3 feet in length. Occasionally, large, loose heads of light yellow flowers appear within the foliage mass; flowers are followed by clusters of bright orange fruits. Slow growth rate; easily transplanted.
GROWING CONDITIONS	Very adaptable; will grow nearly anywhere in Hawai'i, but prefers the hot, dry, nearly desert conditions of the islands' leeward slopes. Requires sunny locations in very well drained soils.
USE	Specimen plant; large container plant; dramatic gray foliage.
PROPAGATION	May be propagated either from seeds or by cuttings.
INSECTS/DISEASES	For scale, apply summer oil or malathion; for mealybugs, diazinon or malathion.
PRUNING	Remove dead and damaged leaves only; old leaves fall naturally from the outer limits of the leaf cluster; they may be removed easily with a light tug. Drastic pruning usually disfigures the plant.
FERTILIZING	Apply general garden fertilizer (10-30-10) to the planting bed at 3-month intervals, and to container plants at monthly intervals.
DISADVANTAGES	None.

104

Dracaena fragrans f. *rothiana*
Fragrant Dracaena

Dracaena fragrans, a very large plant, almost a tree, is native to tropical Africa, especially to Guinea and Sierra Leone. It is the ancestor of several horticultural forms of various foliage types, quite a few of them common in the nursery trade. The green-leaved species itself, although very tropical in appearance, is a rather dowdy dowager in comparison with some of its spectacular offspring. *D. fragrans* f. *rothiana,* for example, has sharply defined white markings running through richly dark green leaves. Both parent and offspring produce large pendent flower clusters which exude a strong fragrance at night.

Although the parent grows in large clumps that easily reach 20 feet in height and diameter, the form *rothiana* is much smaller, more nearly 6 feet. It is favored for both its brilliant foliage and its fragrant flowers. Because it is shade-loving, the plant provides bright color accents in what otherwise might be somewhat gloomy garden areas.

Dracaena, from *drakaina,* meaning female dragon, receives this name because of its close relationship to the dragon's blood trees *(D. cinnabari* and *D. schizantha); fragrans* defines its scent, released at night; *rothiana* honors A. W. Roth (1757–1834), a physician and botanist at Dotlingen, Germany.

HABIT	A woody, evergreen plant that grows to about 6 feet in height, with single or multiple trunks. Dense spiky rosettes of leaves form at tips of trunks and branches. Sharply defined white variegations run through the dark green leaves. These leaves are somewhat more stiff and pointed than are the wavy leaves of the parent. Long pendent stems hold many ball-shaped yellow-orange flower clusters. Orange three-parted fruits follow the bloom. Slow growth rate; easily transplanted.
GROWING CONDITIONS	Quite adaptable but is more tender than most dracaenas; not a beach plant. Prefers rich, well-watered, well-drained soil; requires partial shade and protection from strong winds.
USE	Specimen plant; container plant; colorful tropical foliage.
PROPAGATION	Propagated from cuttings.
INSECTS/DISEASES	To control scale, apply summer oil or malathion. For mealybugs, use diazinon or malathion.
PRUNING	Remove dead and damaged leaves. Drastic pruning of branch tips causes temporarily unsightly stumps, which eventually sprout again. To reduce plant size, remove entire branch at nearest crotch.
FERTILIZING	Apply general garden fertilizer (10-30-10) to the planting bed at 3-month intervals, and to container plants at monthly intervals.
DISADVANTAGES	None.

Dracaena goldieana
Dracaena

Dracaena goldieana's appearance is quite different from that of the other dracaenas grown ornamentally in Hawai'i. Most local dracaenas have rather long, narrow, sword-shaped leaves, while those of this species are quite rounded, being about two-thirds as wide as they are long. Also, the foliage is distinctively marked, showing a similarity to that of its cousin, **bowstring hemp** *(Sansevieria trifasciata).* This dracaena is native to tropical West Africa, where European botanists first collected it about 1860.

It is graceful but dramatic. Its stems are quite slender, yet support spiraled clusters of large leaves. The plant does not send out branches, but several stems may sprout from the roots in much the same way as do the panaxes (*Polyscias* spp.) Its vertical habit allows it to be placed attractively in narrow side yards or in planter boxes. It is also an excellent erect plant for pots.

Dracaena, from *drakaina,* meaning female dragon, the Greek name originally given to the dragon's blood trees *(D. cinnabari* and *D. schizantha); goldieana* is named for this species' first European collector, Rev. Hugh Goldie, of the United Presbyterian Mission in tropical West Africa; Goldie introduced the first plant of this kind to the botanic garden at Edinburgh, Scotland, about 1870.

HABIT A distinctly vertical, woody, evergreen plant that grows to about 8 feet in height. Slender, branchless stems support open rosettes of foliage along their tips. Individual leaves are variegated in shades of green; leaves measure about 5 by 9 inches. Not known to have flowered in Hawai'i, although dense globose clusters of white tubular flowers are produced in its native habitat. Slow growth rate; easily transplanted.

GROWING CONDITIONS Quite adaptable; prefers areas with rich, well-watered, well-drained soil, partial shade (or full sun in the cool valleys), and partial protection from heavy winds. Not a beach plant.

USE Specimen plant; container plant; colorful tropical foliage.

PROPAGATION Generally propagated from cuttings; however, it can be grown from seeds if they are available.

INSECTS/DISEASES To control scale, apply summer oil or malathion. For mealybugs, use diazinon or malathion.

PRUNING Remove dead and damaged leaves. Drastic pruning of stem tips leaves temporarily unsightly stumps, which eventually sprout again; this is the acceptable method to reduce plant height. The pruned leaf crown, with some stem attached, may be planted directly in a new location; with consistent watering, it will develop new roots readily.

FERTILIZING Apply general garden fertilizer (10-30-10) to the planting bed at 3-month intervals, and to container plants at monthly intervals.

DISADVANTAGES None.

Dracaena reflexa cv. 'Song of India'
Dracaena 'Song of India'

This Indian Ocean native, brought to Hawai'i in 1961 by Mrs. A. Lester Marks, is a small, tidy dracaena, very bright and colorful with crisp green and yellow leaves. Its stems are quite slender and pliable, unlike those of many larger relatives; indeed, the plant tends to ''scramble'' and shape itself into Hogarth curving forms. It is a cultivar of the totally green-leafed Mauritian parent, *Dracaena reflexa*, of which many horticultural varieties exist.

The plant is really a shade lover, although it will grow well in places with considerable sunshine. Its bright colors enable it to bring cheerful touches to otherwise dark and overshaded places; and its small size allows it to be set out in narrow, restricted areas. The plant is most satisfactory as a potted plant for shaded lanais and terraces—and even in the interior spaces provided with some window light.

Dracaena, from *drakaina,* meaning female dragon, a term applied to the dragon's blood trees *(D. cinnabari* and *D. schizantha); reflexa,* meaning abruptly bent backward or downward, describes the fountaining foliage.

HABIT
A pliable but woody, bushy, evergreen plant that grows to about 15 feet in height. Older plants exhibit curved, bared stems; leaves striped green and yellow grow in tight rosettes at the stem tips. Leaves are quite pliable, a distinctive characteristic in a genus having many stiff-leafed species. Small erect white flower clusters appear periodically; they are followed by small red-orange fruits. Slow growth rate; easily transplanted.

GROWING CONDITIONS
More tender than the ordinary dracaena species. Prefers very rich, well-watered, well-drained soil and partial to full sunlight. The foliage should be protected from heavy winds and salt air.

USE
Specimen plant; container plant; colorful tropical foliage.

PROPAGATION
Generally is grown from cuttings but may be airlayered.

INSECTS/DISEASES
To control scale, apply summer oil or malathion. For mealybugs, use diazinon or malathion.

PRUNING
Remove only dead and damaged leaves; plant is naturally graceful and its stems should be allowed to grow to their fullest extent.

FERTILIZING
Apply general garden fertilizer (10-30-10) to the planting bed at 3-month intervals, and to container plants at monthly intervals.

DISADVANTAGES
None.

Dracaena sanderiana
Dracaena

Like many of its relatives now growing in Hawai'i, *Dracaena sanderiana* is native to tropical Africa. Its brilliantly variegated and graceful foliage makes it one of the most ornamental members of the genus. It is not so well known to Hawaiian gardeners as are the **money tree** or the **dragon tree.**

This plant is rather diminutive. The leaves are widely spaced along the upright stems, giving the plant an open, airy appearance. The upthrust foliage curves gracefully outward and backward. The plant, distinctly vertical, is an ideal choice for small side yards and planter boxes. It is an excellent container plant for a shaded terrace or lanai, and may be placed inside a room provided with sufficient window lighting.

Dracaena, from *drakaina*, meaning female dragon, a name applied to the dragon's blood trees *(D. cinnabari* and *D. schizantha); sanderiana* honors Henry F. C. Sander (1840–1920), member of a family of nurserymen at St. Albans, England, and Bruges, Belgium.

HABIT
A vertical, woody, evergreen plant that grows to a height of about 6 feet; older plants exhibit straight, bared stems. Narrow green leaves are edged with wide yellow stripes; rosettes of leaves are rather open, each leaf being separated from the others; leaves are not stiff, but bend outward, and are quite pliable. Small, erect, yellowish flower clusters appear periodically; they are followed by small, round, orange-coated fruits. Slow growth rate; easily transplanted.

GROWING CONDITIONS
A tender dracaena, this plant requires protection from excessive sun and wind; it prefers areas of rich, well-watered, well-drained soil; not a dryland or beach plant.

USE
Specimen plant; container plant; terrarium plant; colorful tropical foliage.

PROPAGATION
May be propagated from seeds but generally is grown from cuttings.

INSECTS/DISEASES
To control scale, apply summer oil or malathion. For mealybugs, use diazinon or malathion.

PRUNING
Remove only dead or damaged leaves; pruning should be avoided unless a shortened plant is wanted; in that event, prune tallest stems to nearest crotch or to ground level.

FERTILIZING
Apply general garden fertilizer (10-30-10) to the planting bed at 3-month intervals, and to container plants at monthly intervals.

DISADVANTAGES
None.

Dracaena thalioides
Bremen Ti, Lance Dracaena

Bremen ti, a native of India, was introduced into Hawai'i during the 1920s, when the British liner *Resolute* made frequent stops at Hilo. The ship was equipped with a greenhouse in which this dracaena was growing. During one visit, a part of the plant was given to W. Herbert Shipman, Big Island rancher and plant collector. Somehow, details of the story have become much confused, most especially in the name of the ship. Because the German liner *Bremen* called at Hawaiian ports about that time, people spoke of the plant as the "Bremen ti."

Bremen ti is liked for its handsome big leaves, which are larger than are those borne by most of the dracaenas grown in Hawai'i. It gives somewhat the same visual effect as does the green ti, except that it is much lower and more compact than is its familiar relative. Its stiff, upward-thrusting foliage is arrayed in a partially folded arrangement resembling a fan. The plant is well used in shaded and protected areas where dramatic foliage is wanted. It is an excellent container plant, and will grow undisturbed in the same pot for many years.

Dracaena, from *drakaina*, meaning female dragon, refers to the early Greek name for the dragon's blood trees *(D. cinnabari* and *D. schizantha); thalioides*, meaning like the *Thalia*, refers to the resemblance of this *Dracaena*'s leaves to those of *Thalia dealbata*, a water plant of the arrowroot family (Marantaceae).

HABIT A herbaceous, evergreen plant that grows to about 6 feet in height. Several short, woody stems support long, pleated, oar-shaped, rich green leaves that grow to a foot or more in length; the leaves are attached to the stem by long slender petioles. Insignificant, light cream-colored flowers appear periodically and are followed by small orange fruits. Slow growth rate; easily transplanted.

GROWING CONDITIONS Will not withstand full sun or wind; prefers protected areas with rich, well-watered, well-drained soil and considerable shade.

USE Specimen plant; mass planting; container plant; tropical foliage.

PROPAGATION May be propagated from seeds but generally is grown from root divisions or stem cuttings.

INSECTS/DISEASES To control scale, apply summer oil or malathion. For mealybugs, use diazinon or malathion.

PRUNING Remove dead and damaged leaves; individual crowns may be cut to ground level (each plant generally has several stems and crowns; removing one or two of them ordinarily does not deface it).

FERTILIZING Apply general garden fertilizer (10-30-10) to the planting bed at 3-month intervals, and to container plants at monthly intervals.

DISADVANTAGES None.

Nolina recurvata
Bear Grass, Mexican Tree Lily, Elephant's Foot, Ponytail, Zacate Cortador

The *Nolina*s are natives of the New World. About 30 species from the south-western United States and Mexico are known. Several species that resemble each other share a common name—bear grass; *Nolina recurvata* is also called the Mexican tree lily. The leaves of this species are very long and slender and have sharp, cutting edges, hence the Mexican name zacate cortador, which means cutting grass. Sometimes these leaves are used in Mexico for thatching roofs and for plaiting baskets, mats, and hats. Fiber from the leaves is twisted into rough cordage.

Bear grass is somewhat of a curiosity in Hawai'i: its heavy-footed base (elephant's foot) and fountaining crown (ponytail) cause comment wherever the plant is displayed. It is almost always planted in containers in Hawai'i, even though it grows to tree size in its native habitat. Its slow growth probably has persuaded Islanders to treat it in this way. Thriving as it does in shade or sunlight, the plant is ideal for almost any terrace, lanai, or window-lighted room.

Nolina honors Abbé P. C. Nolin, eighteenth-century French priest (of St.-Marcel Parish), plant collector, author, and landscape architect to Louis XV of France; *recurvata*, meaning curved backward, describes the leaves.

HABIT — An erect, woody, evergreen plant that can grow to more than 30 feet in height in its native habitat but has attained only a fraction of its full growth in Hawai'i. The trunk is distinctively enlarged at its base; older plants may produce branches; both the main trunk and the branches are tipped with graceful rosettes of very narrow, very long leaves (up to 6 feet), which curl downward like a grassy fountain. Insignificant white flower clusters appear occasionally and are followed by small fruits. Slow growth rate; easily transplanted.

GROWING CONDITIONS — Quite adaptable; prefers rich well-watered, well-drained soil; grows happily in sun, filtered shade, or fairly dense shade. Should be protected from strong salt or drying winds.

USE — Specimen plant; container plant; curiosity.

PROPAGATION — Generally propagated from seeds, but may be grown from cuttings.

INSECTS/DISEASES — To control scale, apply summer oil or malathion. For mealybugs, use diazinon or malathion.

PRUNING — Remove dead and damaged foliage only; do not prune plant drastically, else unsightly stumps will result.

FERTILIZING — Apply general garden fertilizer (10-30-10) to the planting bed at 3-month intervals, and to container plants at monthly intervals.

DISADVANTAGES — None.

116

Sansevieria trifasciata var. *laurentii*
Bowstring Hemp, Mother-in-Law's Tongue,
Crocodile's Tongue, Spear Plant, Snake Plant

The *Sansevieria* group is native mainly to South Africa, Madagascar, and southern Arabia. About 60 species are known. As a group, they bear many vernacular names: sometimes they are collectively referred to as bowstring hemp, spear plants, or some other such warrior's name. Malays call the plants crocodile's tongue, in itself an unlikely name, inasmuch as Malay folklore maintains that crocodiles have no tongues. Several tribes refer to the species native to South Africa as snake plants.

The species *S. zeylanica,* from Ceylon, is a chief source of bowstring fiber. Preparations from several others are used to treat intestinal parasites and earache in people living in parts of Africa and Malaysia.

S. trifasciata is one of the world's most popular ornamental plants; several varieties are recognized. Throughout much of the world, the plants are most often grown in containers, but often in Hawai'i, and in places with similar climates, they are set out directly in gardens, where they rapidly form large, spreading colonies. Extremely stiff in habit, the plantings generally are used for dramatic vertical emphasis against architectural features of gardens or residences. The sansevierias are very hardy plants, and will survive much mistreatment.

Sansevieria is named for Raimond de Sansgrio (1710–1771), Prince of Sanseviero, in Italy; *trifasciata,* meaning in three bundles, describes the striated leaves; *laurentii* honors E. Laurent (1861–1904), a botanist in the Belgian Congo.

HABIT A low, stiff, colonial plant that grows to about 3 feet in height; clump is made up of many stiff, upright, leathery, sword-shaped leaves; several leaves arise from the same stem. Periodically the clump blooms, sending up tall flower spikes filled with many small, light yellow-green flowers. Seeds appear only occasionally. Fast growth rate; easily transplanted.

GROWING CONDITIONS Very adaptable, will grow in almost any garden location in Hawai'i, even at the beach. Prefers rich, well-watered, well-drained soil. Will do well in either shady or sunny locations, although plants grown in shade are much darker green.

USE Specimen plant; container plant; mass planting; colorful tropical foliage; long-lasting flower arrangement material.

PROPAGATION Almost always planted from leaf cuttings or root division.

INSECTS/DISEASES To control scale, use summer oil or malathion.

PRUNING Cut leaves back to ground level; new shoots will readily take their place. Do not cut leaves part way to the ground, for resulting stumps are unattractive.

FERTILIZING Apply general garden fertilizer (10-30-10) to the planting bed at 3-month intervals, and to container plants at monthly intervals.

DISADVANTAGES Colonies tend to overgrow allotted garden areas.

Amaryllidaceae
(Amaryllis Family)

The amaryllis family contains 85 genera and about 1,000 species, almost all of which are native to the world's tropical and subtropical regions. Many of the amaryllises are seasonal in that they tend to die down during some part of the year, usually the dry season, then revive after the rains commence. They are often called lilies, but botanists consider them to be quite different from the **Liliaceae.**

The plant from which the family derives its name is the "belladonna lily" *(Amaryllis belladonna),* at present the only species assigned to the genus *Amaryllis.* It is not the source of medicinal belladonna, which comes from *Atropa belladonna* of the nightshade family (Solanaceae).

Other and larger genera in the family include *Crinum, Haemanthus, Hymenocallis, Pancratium, Hippeastrum, Eucharis, Sprekelia, Narcissus,* and *Clivia.* Probably the most popular of the amaryllis relatives are the several horticultural varieties in the genus *Narcissus,* which include the Chinese paper narcissus and the daffodils. The plant group usually known as amaryllis in the nursery trade properly belongs in the genus *Hippeastrum.*

Amaryllis is a Greek girl's name.

Crinum amabile
Sumatran Giant Lily, Spider Lily

Several plants in the amaryllis family have been grouped into an informal—and confused—assemblage known to many people as spider lilies, even though none of them is a lily from the botantists' point of view. With a bit of study, however, these similar yet different plants can be easily identified. The amaryllis with the oldest and clearest claim to the name spider lily is **Hymenocallis littoralis.**

The Sumatran giant lily, *Crinum amabile,* as the common name tells us, is native to the Indonesian island of Sumatra. It is a large plant, very similar to the **Queen Emma lily *(C. augustum),*** but smaller in size and bearing flowers of a paler pink cast than is the case with the Queen Emma lily. The Sumatran giant lily generally is grown in mass arrangements in expansive garden areas, simply because it is a large-scale plant requiring considerable space in which to flourish. Even so, it is the smallest of the three crinums most commonly grown in Hawai'i and, as such, would be the most suitable selection for smaller areas.

Crinum comes from *krinon,* a Greek word meaning lily (which is a prime reason for confusion about the vernacular name for this group of plants); *amabile* means pleasing.

HABIT A herbaceous, evergreen plant with a bulbous base that grows to about 5 feet in height and in diameter; large, succulent, spear-shaped, green-red leaves, 3 to 4 feet long, thrust upward and outward in a sunburst arrangement. Flower clusters appear almost constantly throughout the year; flowers are arranged in a large open nosegay at the tops of broad, succulent stems; individual pink flowers are ''trimmed'' in wine-red. Large green fruits follow the bloom. Fast growth rate; easily transplanted.

GROWING CONDITIONS Very adaptable; will grow in most garden locations; prefers rich, well-watered, well-drained soil in full sun or partial shade. Will withstand considerable salt exposure.

USE Specimen plant; mass planting; dramatic tropical foliage and flowers.

PROPAGATION Easily grown from seeds or from root offshoots.

INSECTS/DISEASES To control spider mites, use wettable sulfur or kelthane.

PRUNING Remove dead and damaged leaves and flower clusters; if drastic pruning is preferred, remove entire plant from the cluster.

FERTILIZING Apply general garden fertilizer (10-30-10) to the planting bed at 3-month intervals.

DISADVANTAGES Subject to attack by spider mites.

Crinum asiaticum
Giant Lily, Spider Lily

The giant lily from tropical Asia is Hawai'i's most common amaryllis, grown nearly everywhere in the Islands. Its puffy fruits, highly salt-tolerant, often are washed out to sea and back again, undamaged. The seeds in the fruits establish themselves readily on sandy beaches, and in some places have become naturalized. The plant is native to tropical Asia, where it has been cultivated since people first settled in those lands. There, the giant lily often is used medicinally; its leaves, oiled and heated, are applied to the body to relieve fevers, headaches, lumbago, or sprains. The Javanese sometimes take the juice of the root internally as an antidote for certain poisons, because it induces vomiting. The root itself is pressed into a poultice for treating wounds.

The giant lily is truly a big plant. Its large bladelike leaves tower over a man's head. Huge yet airy clusters of star-shaped flowers rise majestically above the foliage. The crinums are best used when they are allowed to grow in massive clumps where the great leaves produce dramatic interwoven shapes and shadows. They are ideal beach plants, seeming to be impervious to the most severe salt winds and spray.

Crinum comes from *krinon,* an early Greek name for lilies; *asiaticum* means native to Asia.

HABIT	A herbaceous, evergreen plant, with an extremely bulbous base that grows to 8 feet in height and in diameter; large succulent bright green leaves, 6 feet or more in length, rise above a massive, fleshy stem. Blooms appear almost continuously, producing large clusters of pure white, star-shaped flowers in a sunburst arrangement. Large puffy green fruits follow the bloom. Fast growth rate; easily transplanted.
GROWING CONDITIONS	Exceedingly adaptable; will grow almost anywhere in Hawai'i including beach locations, but attains its greatest size in areas of rich, well-watered, well-drained soil; although the plant grows well in the shade, it prefers hot, dry, sunny areas.
USE	Specimen plant; mass planting; dramatic tropical foliage and flowers.
PROPAGATION	Easily grown from seeds or from root offshoots.
INSECTS/DISEASES	To control spider mites, use wettable sulfur or kelthane.
PRUNING	Remove dead and damaged leaves and flower clusters; if drastic pruning is wanted, remove entire plant from a group.
FERTILIZING	Apply general garden fertilizer (10-30-10) to the planting bed at 3-month intervals.
DISADVANTAGES	Subject to attack by spider mites.

Crinum augustum
Queen Emma Lily

This ornamental, so beloved in Hawai'i, is native to Mauritius and the Seychelles islands in the Indian Ocean. It is quite similar in general appearance to the **Sumatran giant lily, *C. amabile,*** but is much larger and has flowers that are more deeply tinted with red, and its leaves are green, not reddish. The plant was a favorite of Emma Na'ea Rooke, wife of King Kamehameha IV. About 1875, Queen Emma planted some of the first crinums imported to Hawai'i, at Lāwa'ikai, Kaua'i, one of her several residences. Since then it has been known in Hawai'i as the Queen Emma lily.

This is a massive amaryllis that dwarfs a man standing beside it. The enormous sword-shaped leaves seem to be covered with burgundy-colored silk. As with most species of *Crinum,* the Queen Emma lily is best grown in large, dramatic clusters, but great swathes of garden space are required for such mass planting. Fortunately they are excellent for beach sites, being very resistant to salt in air or in soil.

Crinum comes from *krinon,* an early Greek name for lilies; *augustum,* meaning noble, tall, or stately, perfectly describes the plant.

HABIT	A very large, very herbaceous, evergreen plant that grows to 8 feet in height and diameter; large succulent green leaves grow to 6 feet or more in length. The plant is almost constantly in bloom, its light wine-colored flowers being arrayed in sunbursts at the tips of sturdy, wine-red stems. Seeds follow the bloom consistently. Fast growth rate; easily transplanted.
GROWING CONDITIONS	Very adaptable; will grow almost anywhere in Hawai'i, including exposed, salt-sprayed beach areas. Prefers rich, well-watered, well-drained soil. Grows well in extensive shade, but prefers hot, sunny, dry locations.
USE	Specimen plant; mass planting; colorful tropical foliage and flowers.
PROPAGATION	Quickly grown from root offshoots or from seeds.
INSECTS/DISEASES	To control spider mites, use wettable sulfur or kelthane.
PRUNING	Remove dead and damaged leaves and flower clusters; if drastic pruning is wanted, remove entire plant from group; new offshoots will sprout and take the place of the sacrificed plant.
FERTILIZING	Apply general garden fertilizer (10-30-10) to the planting bed at 3-month intervals.
DISADVANTAGES	Subject to frequent attack by spider mites.

Eucharis grandiflora
Amazon Lily, Eucharis Lily

The Amazon lilies belong to the small genus *Eucharis;* all ten of the species are native to tropical South America. They are not true lilies at all, but do have a lily-like appearance that adds to the confusion between the amaryllis and lily families. *E. grandiflora* is rarely used as a garden plant in Hawai'i. People who do cultivate it like it for its short-lived but beautifully pure white flowers. The foliage itself is quite handsome, but the plant is apt to be overlooked when it is not in flower. In the Northern Hemisphere the Amazon lily blooms in the spring, often during the Easter season.

The Amazon lily generally is planted either in containers for shaded lanais or in masses along protected garden borders. Its broad, rich green, somewhat pleated leaves provide a handsome contrast for other more colorful neighbors. It is a good groundcover of moderate height if grown in quantity. During the blooming season it sends forth brilliant white amaryllis flowers, giving much the same impression as do the temperate-climate daffodils and narcissuses.

Eucharis is based on the Greek words meaning very graceful; *grandiflora,* from *grandis,* meaning great, and *flora,* meaning flower, describes this species' large blossoms.

HABIT
A low, clumping, herbaceous, evergreen plant that grows to about 1 foot in height. Many dark green, oar-shaped leaves sprout from bulbous roots; leaves curve outward and backward in a thick fountain-like arrangement. Once each year, during the early spring, clusters of a few handsome amaryllis-shaped white flowers appear on long succulent stems lifted above the leaf mass; fleshy fruits follow the bloom. Moderate growth rate; easily transplanted.

GROWING CONDITIONS
Adaptable; grows extremely well in shaded or partly shaded locations protected from wind, in very rich, well-watered, well-drained soil. Easily damaged by excessive sun or mechanical wear and tear; although sometimes used as a groundcover, the plant will not stand being walked upon.

USE
Specimen plant; container plant; mass planting; spring flowers.

PROPAGATION
Almost always propagated by division, but may be grown from seed.

INSECTS/DISEASES
To control thrips, use diazinon or malathion.

PRUNING
Remove dead and damaged leaves and flower stalks.

FERTILIZING
Apply general garden fertilizer (10-30-10) to the planting bed at 3-month intervals, and to container plants at monthly intervals.

DISADVANTAGES
None.

Haemanthus multiflorus
Blood Lily, Powderpuff Lily

About 50 species of blood lilies are native to tropical and southern Africa and to the island of Socotra in the Indian Ocean. At least two species of *Haemanthus* are used by some African tribes as sources of medicines. In the Cape of Good Hope region, bulbs of *H. coccineus* are stored in vinegar solutions until needed for treatment of asthma and dropsy. The snake lily, *H. natalensis,* has a toxic root which is boiled into a liquid for treating coughs. The blood lily discussed here, however, appears to be purely ornamental.

This plant is much like many of the other amaryllises in that it withers and dies back to ground level during part of the year. It has adjusted to its native southern Africa's dry summer season, December through February. Even when the lily is grown in moist Hawai'i, where conditions resemble those of winter in southern Africa, it dies back as though it were still living there. Because it is a somewhat spectacular oddity, although a beautiful one, people generally grow it in pots, among collections of other interesting and unusual plants.

Haemanthus, from *haema,* meaning blood, and *anthos,* meaning flower, refers to the red flowers of some species; *multiflorus,* meaning many-flowered, describes the powderpuff arrangement of the innumerable tiny florets.

HABIT
A small, herbaceous, deciduous plant that grows to about 1 foot in height. Three or four leaves sprout from the underground bulb; after the growing season, leaves die back to ground level, and plant is dormant for several weeks in the winter. A handsome, rounded powderpuff flower cluster, 4 to 8 inches in diameter, tops a short leafless stem. Flowers, although rather short-lived, are spectacular. Fast growth rate; easily transplanted.

GROWING CONDITIONS
Adaptable; prefers rich, well-watered, well-drained soil; will withstand drought conditions during Hawai'i's winter months when it is dormant. Blooms well in either sunny or partly shaded locations.

USE
Specimen plant; container plant; collector's item.

PROPAGATION
Easily grown from root offshoots or from seeds.

INSECTS/DISEASES
To control thrips, use diazinon or malathion.

PRUNING
Remove withered foliage and flower stems.

FERTILIZING
Apply general garden fertilizer (10-30-10) to the planting bed at 3-month intervals, and to container plants at monthly intervals.

DISADVANTAGES
Dormant during winter and early spring.

Hippeastrum vittatum hybrid
Peruvian Lily, Amaryllis

Members of the genus *Hippeastrum* are what most gardeners think of as amaryllises. They originate in tropical and subtropical America, and about 75 species have been described. *Hippeastrum* species were introduced early into European horticulture; Peruvian lily, *H. vittatum,* a native of that country's Andes, was taken to Europe in 1769. By 1799 hybridization of the imported species had begun; the first hybrid was produced in England. Today almost all ornamental amaryllises are *Hippeastrum* hybrids developed by horticulturists. A widely distributed South American and Caribbean species, the Barbados lily, *H. puniceum,* is also commonly found in Honolulu's gardens.

These are seasonal plants, alternating periods of spectacular floral display with periods of dormancy. Generally the flowers appear from the leafless bulbs, then wither, whereupon the leaves emerge; at the approach of the dormant season, the leaves wither and the plant completes its annual cycle. Gardeners often grow the plants in containers, hidden away for much of the year in secluded places, then exhibit them, full-flowered, on terraces or in borders.

Hippeastrum, from *hippeus,* meaning a knight, and *astron,* meaning a star, is an obscure reference beyond explaining; *vittatum,* meaning striped longitudinally, describes the white flowers striped with red.

HABIT A low, bulbous, herbaceous, and deciduous plant that grows to about 3 feet in height. Tender, strap-shaped leaves emerge from the bulb after the annual flowering period; during the late fall and winter, flowers appear after the leaves have died and the plant has been dormant for several months; flowers rise on a thick, vertical, succulent stalk about 2 feet high; the blooming period lasts a month or so. Some hybrid plants produce seeds; others do not. Fast growth rate; easily transplanted.

GROWING CONDITIONS Quite adaptable because of its seasonal characteristics; requires little attention during its dormant period; gardeners generally tip the pots on their sides and withhold water during those months; pots are righted when blooming season begins and watering is resumed. Flowers need protection from strong, burning sun and wind.

USE Specimen plant; mass planting; container plant; colorful flowers.

PROPAGATION May be propagated by bulb division or from seeds (from which new flower forms are developed).

INSECTS/DISEASES To control thrips, use diazinon or malathion.

PRUNING Remove faded flower clusters; allow leaves to wither before removing; leaves provide nourishment to bulb, and should be allowed to function for as long as possible.

FERTILIZING Apply general garden fertilizer (10-30-10) to the planting bed at 3-month intervals, and to container plants at monthly intervals. Plants need not be fertilized during the dormant period; begin the fertilization program before blooming period.

DISADVANTAGES Dormant for part of the year.

Hymenocallis littoralis
Variegated-leafed Spider Lily, Pancratium

What's in a name? This tropical American amaryllis is the true spider lily. It is seldom called that in Hawai'i, however, because the name is more often associated with the **giant lilies (*Crinum* spp.).** Most often this plant is referred to as pancratium, from its older botanical name. Several forms of spider lily are known. Typically, the species has pure green leaves, but the plant pictured here is one of the parents' more popular ornamental offspring. The bulbs are said to be poisonous; those of a Caribbean relative, the white lily *(H. speciosum),* are used by Carib Indians as the source of an emetic.

Spider lilies are extremely hardy. They are among the best of Hawai'i's salt-resistant seashore plants, as the specific name, *littoralis,* suggests. This variety of spider lily is a small plant, much smaller than are the crinums: it has short strap-shaped leaves similar to those of the **amaryllis hybrids (*Hippeastrum).*** Spider lilies generally are grown in masses as flowering groundcover or border plants. They are excellent for containers, also.

Hymenocallis, from *hymen,* meaning membrane, and *kallos,* meaning beautiful, refers to the handsome, thin tissue connecting the stamens; *littoralis,* meaning of the shore, refers to the species' natural habitat.

HABIT | A low, clumping, herbaceous, evergreen plant that grows to about 2 feet in height. The two most popular garden types have leaves that are either pure green or variegated green and cream. Small clusters of distinctive white flowers, 6 inches across, appear almost constantly. The star-shaped flowers are easily distinguished from those of the crinums because of the membranous tissue connecting the petals at their bases, somewhat like the webbing on ducks' feet. Crinums do not have this webbing. Occasionally seeds appear on the maturing flower stalks. Fast growth rate; easily transplanted.

GROWING CONDITIONS | Very adaptable; will grow nearly everywhere in Hawai'i; an excellent beach plant; grows in almost any garden soil including the very sandy. Blooms best in open, sunny areas, but the form with variegated leaves adds bright color to dark areas.

USE | Specimen plant; mass planting; container plant; interesting, constant flowers and colorful tropical leaves.

PROPAGATION | By division of clumps.

INSECTS/DISEASES | To control thrips, apply diazinon or malathion. For spider mites, use wettable sulfur or kelthane.

PRUNING | Remove old flower clusters and dead foliage only; the plant is quite tidy.

FERTILIZING | Apply general garden fertilizer (10-30-10) to the planting bed at 3-month intervals, and to container plants at monthly intervals.

DISADVANTAGES | Bulbs may be poisonous.

Sprekelia formosissima
Jacobean Lily, St. James Lily, Flor de Mayo, Pata de Gallo, Panānā

This native of Mexico is a plant of many names. English-speaking peoples generally call it the Jacobean lily, or, more rarely, the St. James lily. Spanish-speaking peoples know it as flor de mayo (May flower) or pata de gallo (cock's foot). Hawaiians named it in honor of the chiefess Pānānā, wife of Samuel Parker, a son of the man who founded the Big Island's enormous cattle ranch.

The plant is a potent medicinal: it contains the alkaloid amaryllin, a poisonous substance used by Mexicans to induce vomiting. The preparation is known as *amarilina* in Mexico.

The Jacobean lily flowers only fleetingly. It is dormant for a considerable part of the year, but during its winter-to-summer growing season, leaves and flowers appear. The foliage may be somewhat inconspicuous, but the flowers are spectacular. They are great velvety crimson blossoms that blaze in splendor for a few days, then just as quickly disappear.

Sprekelia is named for J. H. von Sprekelsen (d. 1764), a botanist of Hamburg, Germany, who sent bulbs of this plant to Linnaeus, the great Swedish botanist; *formosissimus*, meaning the most beautiful, accurately describes the flowers. The plant was introduced to Europe, by way of Spain, in the late 16th Century, and reached England in 1658.

HABIT — A low, herbaceous, deciduous plant that grows to about 3 feet in height. Dormant during the summer months; in the fall leaves and flowers appear, but blossoms for only a short period. Flowers are quite large, 5 to 6 inches in diameter; they are brilliantly red and have a velvety surface texture. Flower stalks appear one after the other, on opposite sides of the bulb. Fast growth rate; easily transplanted.

GROWING CONDITIONS — Plants grow best in Hawai'i's cooler regions, but do fairly well at lower elevations. Require rich, well-watered, well-drained soil and full sun or partial shade.

USE — Specimen plant; container plant; cut flower material.

PROPAGATION — Propagated by separating the bulb offsets during the dormant season.

INSECTS/DISEASES — To control thrips, use diazinon or malathion.

PRUNING — Remove dead flower stalks; leaves will wither naturally; allow them to remain on the plant until they are quite dry, inasmuch as they provide nourishment to bulbs for the following season.

FERTILIZING — Apply general garden fertilizer (10-30-10) to the planting bed at 3-month intervals during the growing season and just before the blooming season.

DISADVANTAGES — Dormant for part of the year. Bulbs are poisonous.

Taccaceae
(Tacca Family)

The tacca family is represented by two genera: *Tacca,* which has about 30 recognized species, and a genus from China, *Schizocapsa,* with but a single species. All of the taccas are tropical plants, originating mainly in Southeast Asia, although a few are American. Several are useful. *T. fatsiifolia,* native to the Philippines and Indonesia, is used in those regions for the treatment of snakebites and wounds. *T. palmata,* from the same islands and the Malay peninsula, has roots that are chewed or prepared in a tonic to relieve stomach complaints. *T. artocarpifolia,* from Madagascar, has edible tubers.

The most important species in Polynesia is the arrowroot, *T. leontopetaloides.* Pia, as it is known to Hawaiians, was an important source of starch for the Islanders. After separating the starch from the tubers, Hawaiians added it to coconut milk and then either baked or heated the mixture until it set. The result was the delicious pudding called *haupia.* Today's confection by that name generally employs corn starch as the thickening ingredient.

Taccaceae receives its name from *tacca,* a Malayan word for several related species.

Tacca chantrieri
Bat Flower

The bat flower, so called because of its distinctively shaped, purplish black flowers, is native to Thailand and Burma. It is an ornamental relative of Polynesian arrowroot, or pia, *Tacca leontopetaloides*. Little practical use is made of this plant, although several of its relatives produce tubers that are valued either as food or as medicine. Unusual as are the bat flower's shape and color, it is but one of several distinctively flowered relatives. Other species bear flowers that are green and brown, or green with yellow and purple variegation, or green and violet, to mention only a few examples.

This is a jungle plant: it grows best in shaded garden retreats, where its large leaves are protected from damage by wind, rain, or falling debris. The flowers are striking: once seen, they are not forgotten. The plant itself makes a handsome high groundcover or walkway border. It also grows well in containers, being especially suited to protected lanais.

Tacca is the vernacular name for a number of species native to the Malaysian archipelago-peninsula; *chantrieri* honors Chantrier Frères, well-known early nurserymen of Mortefontaine, France.

HABIT A low, herbaceous, evergreen plant that grows to 2 feet in height, 3 feet in diameter. Many leaves appear in a loose rosette from tuberous roots; individual leaves, about 9 by 20 inches, are large and handsome. Periodically the purplish-brown, almost black, flower clusters rise among the bright green leaves; the effect is stunning. Occasionally seeds appear after the flowers fade. Fast growth rate; easily transplanted.

GROWING CONDITIONS Requires a great deal of moisture and protection from sun and wind; grows best in humus-rich, well-watered, well-drained soil, in deep or partial shade. Large leaves must be protected from damage; plants should not be placed under trees or shrubs which scatter heavy debris.

USE Specimen plant; mass planting; container plant; tropical foliage and unusual flowers.

PROPAGATION By root division.

INSECTS/DISEASES To control thrips, use diazinon or malathion.

PRUNING Remove dead and damaged leaves and flowers only; with adequate protection, the plant will maintain its beautiful natural lines.

FERTILIZING Apply general garden fertilizer (10-30-10) to the planting bed at 3-month intervals, and to container plants at monthly intervals.

DISADVANTAGES Requires more protection than do most garden plants.

Musaceae
(Banana Family)

The edible bananas are among the oldest of all the world's food crops. Their agricultural beginnings, the details of which are lost in the dim recesses of history, precede by far most of the grains, vegetables, and fruits eaten today. Like all agricultural crops, the bananas have been developed through the ages-old process of cultivation and selection for improved varieties. The important edible species are the common banana *(Musa acuminata × balbisiana)*, the Chinese banana *(Musa × nana)*, the plantain *(Musa × paradisiaca* var. *normalis)*, and the fehi of Tahiti *(Musa troglodytarum* var. *acutibracteata)*. Apparently most of these edible species originated in tropical Asia and were dispersed from there to the rest of the tropics by early migrants from that continental center.

Almost all parts of the plant can be used in some way. Species bearing inedible fruits (or none at all) provide materials for shelter, clothing, matting, and packing. *M. textilis,* from tropical Asia, has fibers that are woven into the familiar abaca fabric; *M. basjoo,* from Japan, produces basho fabric and fine papers for calligraphers to write upon.

The banana family consists of two genera, *Musa* and *Ensete.* Forty-two species in the two genera are recognized, and untold numbers of cultivars have been developed. Formerly, plants such as the heliconias and the birds of paradise were considered to be members of the banana family. Now those plants are assigned to their own families, **Heliconiaceae** and **Strelitziaceae.** The family Musaceae is named in honor of Antonius Musa, a physician to Octavius Augustus, first emperor of Rome, 63 B.C.–A.D. 14.

143

Musa coccinea
Red-flowering Banana

The red-flowering banana is an ornamental relative of the many utilitarian species. It is native to tropical Southeast Asia, especially Burma, Cambodia, and Viet Nam. Because its gorgeous flowers last for many days, they are used in decorative arrangements for homes or for religious or processional occasions. Often the brilliant scarlet flowers are arranged with assorted fruits and rice cakes in gigantic cones prepared for special village and temple observances.

The plant itself differs to some extent from those of the common edible bananas. It is much smaller and more slender, and its flowers grow upright, emerging from the crown on long sturdy stalks, whereas flowers of most edible bananas hang in clusters down the side of the tree. Because its striking flowers are so very different from those of the better-known eating bananas, many people mistake it for a kind of ginger plant. In the garden, the red-flowering banana is planted in much the same way as are heliconias and gingers, in large clumps for borders and as backgrounds. The flowers are seasonal, but the handsome foliage is continually attractive if it is protected from wind and falling debris.

Musa is named for Antonius Musa, a physician to the Emperor Augustus of Rome (63 B.C.–A.D. 14); *coccinea,* meaning scarlet, describes the vivid flower bracts.

HABIT
An upright, herbaceous, evergreen plant that grows to about 12 feet in height; in the typical clumping banana habit, the individual stalks sprout from roots, grow to maturity, flower, then die; as the old stalk dies, new stalks take its place. Large, bright green, typical banana leaves. Flowers are about 8 inches long and 6 inches in diameter; the flower consists of brilliant orange-red bracts that surround insignificant yellow flowers. Small, inedible bananas form on the stalk after the blooming period. Fast growth rate; easily transplanted.

GROWING CONDITIONS
Requires a certain amount of protection; does not do well in exposed locations where extensive sun and wind burn the foliage and flowers; not a beach plant. Prefers moist jungle conditions and rich, well-watered, well-drained soils. If perfect leaves are wanted, they must be protected from falling debris.

USE
Specimen clump; mass planting; tropical foliage and handsome flowers.

PROPAGATION
By root division.

INSECTS/DISEASES
To control thrips, use diazinon or malathion.

PRUNING
Remove old and damaged leaves; cut stalks of plants (and flowers in season) at ground level; new stalks will take the former's place.

FERTILIZING
Apply general garden fertilizer (10-30-10) to the planting bed at 3-month intervals.

DISADVANTAGES
None.

144

Musa sumatrana
Sumatran Banana

This native of Indonesia is one of the most beautiful of the banana species. It is much favored in the tropics, both as a garden plant and as a decoration for special occasions. Sometimes whole plants are used in mammoth arrangements combined with various tropical flowers. Also, as in Hawai'i, the leaves are laid flat on tables, to become cool and attractive (and disposable) serving surfaces.

The Sumatran banana produces neither attractive flowers nor edible fruit. But it compensates for those deficiencies in its exquisite foliage: each leaf is different from the others, looking as if it has been hand-painted by an expert in the use of lacquer. The large leaves, whose tips may reach well above the roof lines of most tropical residences, are massive in scale and dramatic in effect. This is decidedly a tropical plant, both in habit and in use; when carefully grown, it produces some of the most perfect foliage ever to be seen in tropical gardens. As with all bananas, however, the lovely leaves require complete protection from destructive winds and branches of neighboring trees.

Musa is named for Antonius Musa, physician to Octavius Augustus, first emperor of Rome (63 B.C.–A.D. 14); *sumatrana* indicates the island where this species was discovered by Western botanists.

HABIT A large, tall, herbaceous, evergreen plant that grows to about 20 feet in height; the plant, as with most species of *Musa,* sprouts from an underground stem, grows to maturity, flowers, then dies; as the old plant dies, new stalks take its place. Large "trunks," about 8 feet tall, support several beautifully variegated leaves; leaves, which may reach 10 to 12 feet in length, have a moss-green ground decked with splotches of cinnabar red. Insignificant flowers and fruits. Fast growth rate; easily transplanted.

GROWING CONDITIONS Requires much protection if intact foliage is wanted; best planted in a jungle-like environment, in the lee of a considerable windbreak. Grows well in full sun or partial shade; prefers rich, well-watered, well-drained soil. Requires considerable space in which to spread.

USE Specimen plant; mass planting; colorful tropical foliage.

PROPAGATION By root division.

INSECTS/DISEASES To control thrips, use diazinon or malathion.

PRUNING Remove old and damaged leaf stalks at ground level. Individual fresh leaves may be removed easily for flower arrangements; in this event, care should be taken not to injure or remove the stalk's growing tip.

FERTILIZING Apply general garden fertilizer (10-30-10) to the planting bed at 3-month intervals.

DISADVANTAGES None of any consequence.

Musa velutina
Pink-fruited Banana, Self-peeling Banana

Bananas are fundamental to the daily lives of people in Asia: they provide both food and materials for shelter, and they give sustenance to the human spirit as well. In many native mythologies, the banana is the tree of life and rejuvenation, because, as old plants bring forth fruits and then die, so do the new shoots sprout and grow. Many references to bananas can be found in Buddhist and Hindu sacred literature, and the banana leaf itself symbolizes the entire Buddhist scriptures. To Hindus it is the tree of knowledge, under which a scholar may sit to gain intellectual nourishment. The Chinese hold the view that the banana symbolizes self-improvement through education. And some mythographers maintain that the fruit with which the Serpent tempted Adam and Eve in the Garden of Eden was a banana, not an apple.

This *Musa* is another one of several species that produce inedible but very ornamental bananas. In South and Southeast Asia, the decorative fruits and flowers are used in ceremonial arrangements celebrating religious and secular occasions. It is a small plant, a dwarf in comparison to most species of banana. The fruits have the interesting characteristic of peeling themselves; as they ripen, the peels gradually fold back, exposing the creamy flesh filled with seeds.

Musa is named for Antonius Musa, a physician to Octavius Augustus, first emperor of Rome (63 B.C.–A.D. 14); *velutina*, meaning covered with down, describes the velvety fruits.

HABIT A small, erect, herbaceous, evergreen plant that grows to about 6 feet in height; stalks are slender, 2 to 3 inches in diameter, each formed by several banana leaves; erect pink flower clusters appear among the leaves, one cluster to each stalk, forming brilliantly pink, velvety fruits. Fruits ''peel themselves'' when ripe. Fast growth rate; easily transplanted.

GROWING CONDITIONS Very adaptable; will grow in many Hawaiian locations but not in dry land or places exposed to beach winds; the plant's small size helps protect it from wind damage. Prefers rich, well-watered, well-drained soil and full sun or partial shade.

USE Specimen plant; mass planting; tropical foliage and attractive flowers and fruit; flower arrangement material.

PROPAGATION Generally by root division, but may be grown from seed. Root division is the faster method.

INSECTS/DISEASES To control thrips, apply diazinon or malathion.

PRUNING Remove old and damaged leaf stalks at ground level; individual leaves, flowers, and fruits may be removed without harm, as the entire stalk withers when the fruits have ripened.

FERTILIZING Apply general garden fertilizer (10-30-10) to the planting bed at 3-month intervals.

DISADVANTAGES None.

Heliconiaceae
(Heliconia Family)

The heliconia family is tropical American in origin. About 80 species are recognized, all of which are placed in the single genus, *Heliconia*. In the Caribbean region they are known collectively as wild plantains or wild bananas. Botanists at one time considered these plants to be part of the banana family. In general the heliconias are grown as garden ornaments and for floral material. A few species are used in other ways as well. *H. bihai* has edible young shoots, and sometimes its leaves and those of *H. schiedeana* are employed for thatching roofs in Mexico and in parts of the Caribbean. *H. brasiliensis* has roots and seeds that are used medicinally in Brazil.

Heliconia species range in size from 3-foot dwarfs to tree-sized (30-foot) giants. Many kinds of flower forms are known; most species produce fairly distinctive flower clusters. The colorful, seemingly sculptured bracts are the most visible parts that give the flower clusters their "exotic" and "tropical" appearance. The heliconias were named for Mount Helicon in Greece, the abode of the Muses, the nine goddesses of the arts and sciences in Greek mythology.

Heliconia aurantiaca
Dwarf Golden Heliconia, Platanillo

This pert plant is one of the smaller ornamental heliconias. It is a native of Mexico, as are many of its closest relatives. Spanish-speaking peoples know these plants as platanillo, or little plantain, because their stems and leaves are much like those of the banana in appearance.

The dwarf golden heliconia is generally grown in masses, often beneath true bananas or other and larger kinds of heliconias. This species produces flowers that are excellent cut material for arrangements. The flowers are somewhat smaller than are those of most other ornamental heliconias and blend well in mixtures of tropical blossoms. Those shown here were photographed in the bud stage. As a general rule, heliconias require considerable protection from wind and falling debris, because their large, beautiful leaves are quite easily torn.

Heliconia is named for Mount Helicon in Greece; *aurantiaca*, meaning orange, refers to the color of the flower bracts. Sometimes the heliconias as a group are called false birds of paradise, because of their resemblance to those closely allied plants in the family **Strelitziaceae.**

HABIT	A small, herbaceous, evergreen plant that grows to about 30 inches in height, in a distinctly clumping habit, with several leaf stalks sprouting from each root mass; the leaves are quite like those of the bananas, only much smaller reaching about 10 inches in length. Colorful orange bracts enclose insignificant red and green flowers. Old foliage dies back during winter months. Fast growth rate; easily transplanted.
GROWING CONDITIONS	Grows best in protected environments in humus-rich, well-watered, well-drained soils; blooms well in areas of full sun or partial shade.
USE	Mass planting; planter boxes; colorful flower clusters and tropical foliage; cut flower material.
PROPAGATION	Generally by root division but may be grown from seed.
INSECTS/DISEASES	To control scale, use malathion.
PRUNING	Remove dead leaf and flower stalks after blossoming; generally the entire clump is cut to the ground after the flowering period is completed. New foliage grows back readily.
FERTILIZING	Apply general garden fertilizer (10-30-10) to the planting bed at 3-month intervals. In poor soils plant shows evidence of deficiencies in minor elements (yellow leaves); apply fertilizers containing minor elements at 3-month intervals to correct this condition.
DISADVANTAGES	None.

Heliconia caribaea
Giant Yellow Heliconia

The giant yellow heliconia, a relatively new introduction, is one of Hawai'i's most spectacular ornamentals. A native of the Caribbean islands, it was brought to Hawai'i in 1958 by Mr. and Mrs. W. W. Goodale Moir of Honolulu. Caribbean people eat the young shoots. They also use the leaves to cover foods while these are being cooked in underground ovens; the foliage imparts a pleasing taste to the food being baked. In Hawai'i, the huge flower clusters are grown for the florist trade.

The plant is extremely large; its bulk approaches or exceeds that of some bananas. Its handsome banana-like leaves are grand in scale, blemish-free, and give a splendidly tropical effect when grown in protected gardens. The sturdy, waxy, yellow flower clusters grace the plant for long periods; when cut they are dramatic components in large exotic flower arrangements.

Heliconia is named for Mount Helicon in Greece; *caribaea* indicates this species' native habitat.

HABIT   A large, upright, herbaceous, evergreen plant that grows to about 20 feet in height. Several long, sturdy leaf stalks sprout from the underground rhizomes; large, banana-like leaves are borne toward the stalk tops. Flower clusters appear mostly from late spring to late fall; clusters are very large, growing to about 8 inches wide and 12 inches long; bracts are a bright, waxy yellow; insignificant flowers nestle within the bracts. Old foliage dies back during winter. Fast growth rate; easily transplanted. Some forms of this plant exhibit red or maroon bracts rather than the more commonly seen yellow.

GROWING CONDITIONS   Quite adaptable, but grows best in humus-rich, well-watered, well-drained soil; banana-like leaves must be protected from wind and falling debris. Blooms well in areas of full sun or partial shade. Not a beach plant, but will grow near the beach if carefully protected from salt spray and winds.

USE   Specimen plant; mass planting; tropical foliage and flowers; excellent cut flower material.

PROPAGATION   Generally by root division but may be grown from seeds.

INSECTS/DISEASES   To control scale, use malathion.

PRUNING   Remove dead leaf and flower stalks after flowering; generally the entire clump is cut to the ground after flowering period is completed. New foliage grows back readily.

FERTILIZING   Apply general garden fertilizer (10-30-10) to the planting bed at 3-month intervals. Yellow, sickly leaves indicate deficiencies in minor elements, especially iron, in poor soils; apply minor element fertilizer to the planting bed at 3-month intervals.

DISADVANTAGES   Established plantings may outgrow garden area allotted to them.

Heliconia humilis
Lobster's Claw, Platanillo

The lobster's claw is well named, for this shrubby plant from tropical America produces spectacular flower clusters with crimson bracts that resemble exactly a string of boiled crustacean pincers. Spanish-speaking peoples know this heliconia as platanillo, or little plantain, because the large oval leaves are so similar to those of the edible plantains. This platanillo is almost entirely ornamental, although sometimes its small, blemish-free leaves are used for covering foodstuffs in underground ovens and as serving plates. In Hawai'i and other tropical areas the plants are grown commercially for the florist trade.

Heliconia humilis is one of the smaller ornamental members of the genus. Ordinarily it is grown in large masses with other heliconias, the ornamental or fruiting bananas, or similar large-leafed plants. Flowers, produced throughout much of the year's warm season, provide good material for tropical arrangements. As with all bananas and heliconias, once the flower cluster has matured, the entire leaf stalk withers and dies; for this reason, gardeners generally cut the stalk to the ground when they gather the flowers.

Heliconia is named for Mount Helicon in Greece, the legendary mountain of the Muses; *humilis,* meaning low, describes this species' low growth habit as compared with the other, larger heliconias.

HABIT | An upright, herbaceous, evergreen plant that grows to about 6 feet in height. Several sturdy leaf stalks sprout from the underground rhizomes; small banana-like leaves are borne at the stalk tops. Flowers appear mostly between late spring and late fall; upright flower clusters are composed of brilliant red bracts arranged alternately on the stem; insignificant white flowers nestle within the bracts. Old foliage dies back during winter. Fast growth rate; easily transplanted.

GROWING CONDITIONS | Quite adaptable, but grows best in humus-rich, well-watered, well-drained soil; banana-like leaves must be protected from wind and falling debris. Blooms well in areas of full sun or partial shade. Not a beach plant, but will grow near the beach if well protected from salt spray and winds.

USE | Specimen plant; mass planting; tropical foliage and flowers; excellent cut flower material.

PROPAGATION | Generally by root division, but may be grown from seed.

INSECTS/DISEASES | To control scale, use malathion.

PRUNING | Remove dead leaf and flower stalks after flowering; generally the entire clump is cut to the ground after the flowering period is over. New foliage grows back readily.

FERTILIZING | Apply general garden fertilizer (10-30-10) to the planting bed at 3-month intervals. Yellow, sickly leaves indicate deficiencies in minor elements, especially iron, in poor soils; apply minor element fertilizer to the planting bed at 3-month intervals.

DISADVANTAGES | Established plantings may outgrow garden area allotted to them.

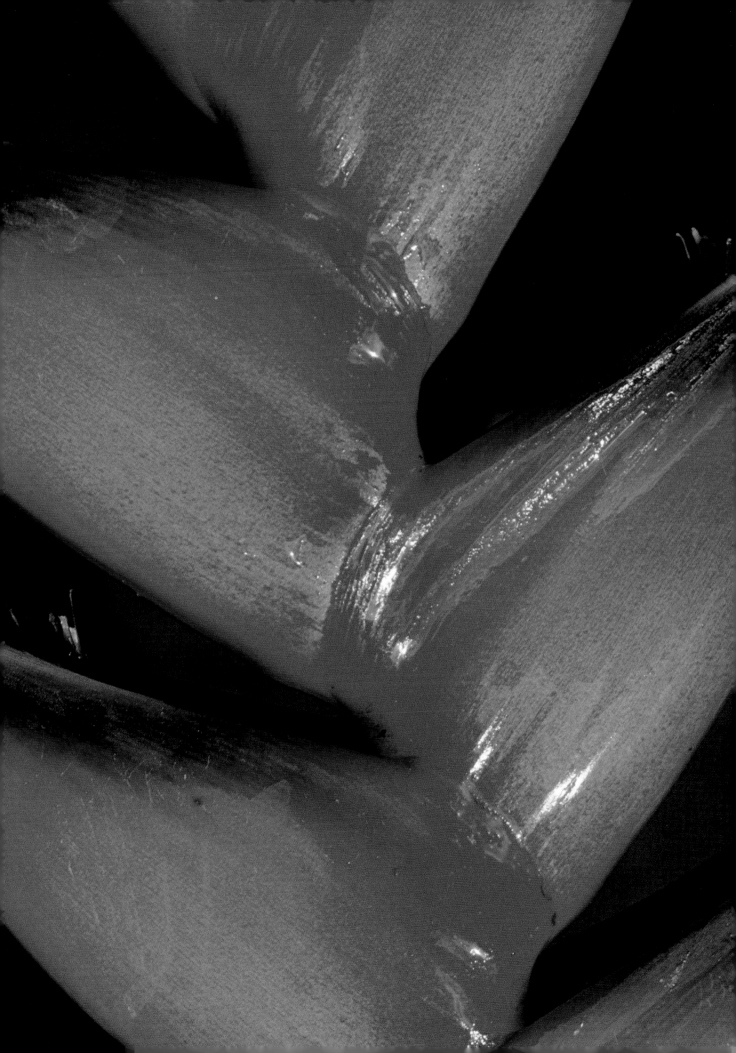

Heliconia sp. cv. 'Dwarf Humilis'
Jamaican Heliconia, Dwarf Humilis Heliconia

This *Heliconia* is a relative newcomer among ornamental plants. It is a native of the Caribbean region, most probably from Trinidad and Jamaica, and is still an undescribed species. The appellation 'Dwarf Humilis' is an unofficial trade name used by plant collectors. Hiroshi Tagami and Richard Hart introduced this cultivar to Hawai'i in 1971. Its flowers are somewhat similar to those of **H. humilis,** but the two plants should not be confused with each other.

This plant is considerably smaller than even the dwarfish *H. humilis.* It grows to little more than knee height. Currently somewhat scarce in commercial nurseries, eventually this plant should be used extensively in masses as border material in conjunction with other heliconias and similar herbaceous plants. Its low growth also makes it valuable for groundcover and foreground plantings.

Heliconia is named for Mount Helicon in Greece; its descriptive specific name has yet to be determined.

HABIT
An upright, herbaceous, evergreen plant that grows to about 2 feet in height. Several sturdy leaf stalks sprout from the underground rhizomes; small banana-like leaves appear at the stalk tops. Flowers develop mostly between late spring and late fall; clusters are composed of several loosely arranged red bracts containing small white flowers. Old foliage dies back during winter. Fast growth rate; easily transplanted.

GROWING CONDITIONS
Quite adaptable, but grows best in humus-rich, well-watered, well-drained soil; banana-like leaves must be protected from wind and falling debris. Blooms well in areas of full sun or partial shade. Not a beach plant, but will grow near the sea if well protected from salt spray and winds.

USE
Specimen plant; mass planting; tropical foliage and flowers; excellent cut flower material.

PROPAGATION
Generally by root division, but may be grown from seed.

INSECTS/DISEASES
To control scale, use malathion.

PRUNING
Remove dead leaf and flower stalks after blossoming; generally the entire clump is cut to the ground after the flowering period is completed. New foliage grows back readily.

FERTILIZING
Apply general garden fertilizer (10-30-10) to the planting bed at 3-month intervals. Yellow, sickly leaves indicate deficiencies in minor elements, usually iron, in poor soils; apply minor element fertilizer to the planting bed at 3-month intervals.

DISADVANTAGES
Established plantings may outgrow garden area allotted to them.

158

Heliconia psittacorum
Parrot's Beak Heliconia

The parrot's beak probably is the heliconia most commonly grown in Hawai'i. It is native to the islands of the Caribbean and grows equally well in Hawai'i's similar environment. Strictly ornamental, both here and in the Caribbean, it is one of the smaller *Heliconia* species, reaching to about hip height. It grows readily under many conditions and has become naturalized in some areas, where it is practically a weed. The flowers, almost constantly in bloom, add much color to local gardens. The flower clusters are often used in flower arrangements. It was introduced to Hawai'i from Puerto Rico in 1950 by Robert and John Gregg Allerton of Lāwa'i, Kaua'i.

Generally this species is grown as a border plant or as a mass groundcover in association with other, taller, heliconias and bananas. Its small size and sturdy flowers allow it to be planted in more exposed garden locations than its larger and more tender leafed relatives can tolerate.

Heliconia is named for Greece's Mount Helicon, the legendary home of the Muses; *psittacorum,* meaning of the parrot, describes the tiny, black and orange flower bracts shaped like the beak of a parrot.

HABIT An upright, herbaceous, evergreen plant that grows to about 5 feet in height. Many stiff, slender leaf stalks sprout from underground rhizomes; small, banana-like leaves appear at the tops of the stalks. Blooms plentifully during spring months, then sporadically until late fall; flower clusters are miniature by *Heliconia* standards; the entire inflorescence is about 5 inches in diameter; clusters consist of bright red bracts and black-tipped orange flowers. Old foliage dies back during winter. Fast growth rate; easily transplanted.

GROWING CONDITIONS Quite adaptable, but grows best in humus-rich, well-watered, well-drained soil; banana-like leaves must be protected from wind and falling debris. Blooms well in areas of full sun or partial shade. Not a beach plant, but will grow near the sea if well protected from salt spray and winds.

USE Specimen plant; mass planting; tropical foliage and flowers; excellent cut flower material.

PROPAGATION Generally by root division, but may be grown from seed.

INSECTS/DISEASES To control scale, use malathion.

PRUNING Remove dead leaf and flower stalks after flowering; generally the entire clump is cut to the ground after the blossoming period is completed. New foliage grows back readily.

FERTILIZING Apply general garden fertilizer (10-30-10) to the planting bed at 3-month intervals. Yellow, sickly leaves indicate deficiencies in minor elements, especially iron, in poor soils; apply minor element fertilizer to the planting bed at 3-month intervals.

DISADVANTAGES Established plantings may outgrow garden areas allotted to them.

Heliconia roseo-striata
Pinkstripe-Leaf Heliconia

This beautiful foliage plant stands out even in a family noted for its exotic tropical flowers. The colorful pink, green, and red striations within the enormous leaves remind one of delicate feathers observed through a magnifying glass. The plant is almost entirely decorative, although in some places in its native tropical America the leaves are used in the preparation and serving of foods. Several other foliage heliconias are used in similar decorative fashion; these include *Heliconia illustris,* which has green upper-leaf surfaces veined with red and copper-red under-leaf surfaces; *H. indica,* whose several color forms have variations of green and cream or pink leaf stripes; and *H. spectabilis,* which has finely striped green and red leaves.

The pinkstripe-leaf heliconia is a large plant, massive in scale and dramatic in color. Generally it is seen as a specimen plant in residential gardens. This heliconia is a fine selection for large, protected planter boxes or sheltered terraces. Its splendid foliage works well with tall, massive architectural forms. The plant's leaves must be protected from wind and debris falling from trees.

Heliconia is named for Mount Helicon of Greece; *roseo-striata,* from *roseus,* meaning rose colored, and *striata,* meaning marked with fine parallel lines, describes the handsome foliage.

HABIT
: A tall, upright, herbaceous, evergreen plant that grows to about 12 feet in height. Several long sturdy leaf stalks sprout from the underground rhizomes; large handsomely colored banana-like leaves appear toward the stalk tops. Old foliage dies back during winter. Fast growth rate; easily transplanted.

GROWING CONDITIONS
: Quite adaptable, but grows best in humus-rich, well-watered, well-drained soil; banana-like leaves must be protected from wind and falling debris. Grows well in areas of full sun or partial shade. Not a beach plant, but will grow near the sea if well protected from salt spray and winds.

USE
: Specimen plant; mass planting; tropical foliage and flowers; excellent material for arrangements.

PROPAGATION
: Generally by root division, but may be grown from seeds.

INSECTS/DISEASES
: To control scale, use malathion.

PRUNING
: Remove dead leaf and flower stalks after flowering; generally the entire clump is cut to the ground after the blossoming period is completed. New foliage grows back readily.

FERTILIZING
: Apply general garden fertilizer (10-30-10) to the planting bed at 3-month intervals. Yellow, sickly leaves indicate deficiencies in minor elements, especially iron, in poor soils; apply minor element fertilizer to the planting bed at 3-month intervals.

DISADVANTAGES
: Established plantings may outgrow garden areas allotted to them.

Heliconia rostrata
Dwarf Hanging Heliconia

This native of Peru is one of several *Heliconia* species whose flower clusters hang downward. Among the hanging heliconias being grown in Hawai'i are *H. collinsiana* and *H. nutans,* whose waxy flower spathes are red, and *H. platystachys,* whose multicolored spathes are similar to those of this dwarf species. These heliconias are grown mostly for ornament, although flowers are produced commercially both in Hawai'i and in other tropical regions. The long-lasting, pendent flowers are favorite materials for including in large, "exotic" arrangements.

Although it is called a dwarf heliconia, this plant does grow higher than a man. It is a much larger plant than **H. aurantiaca,** for example, or than **H. humilis** and the **'dwarf humilis.'** It gained this name only because it is a dwarf by comparison with most species of hanging heliconias. The plant is best grown in masses and displayed in a jungly sort of background. Because its foliage quickly shows damage from wind and normal processes of aging, the plant is grown mostly for its durable flowers.

Heliconia is named for Mount Helicon in Greece; *rostrata,* meaning ending gradually in a long, straight, hard point, describes the flower bracts' yellow terminals.

HABIT An upright, herbaceous, evergreen plant that grows to about 8 feet in height; several sturdy leaf stalks sprout from underground rhizomes; large banana-like leaves appear toward the tops of the stalks. Flowers are borne mostly between late spring and late fall; flower clusters, distinctively pendent, are composed of yellow-tipped red bracts which enclose inconspicuous yellow flowers. Old leaf stalks die back during winter months. Fast growth rate; easily transplanted.

GROWING CONDITIONS Quite adaptable, but grows best in humus-rich, well-watered, well-drained soil; banana-like leaves must be protected from wind and falling debris. Blooms well in areas of full sun or partial shade. Not a beach plant, but will grow near the sea if well protected from salt spray and winds.

USE Specimen plant; mass planting; tropical foliage and flowers; excellent cut flower material.

PROPAGATION Generally by root division, but may be grown from seed.

INSECTS/DISEASES To control scale, use malathion.

PRUNING Remove dead leaf and flower stalks after blossoming; generally the entire clump is cut to the ground after the flowering period is completed. New foliage grows back readily.

FERTILIZING Apply general garden fertilizer (10-30-10) to the planting bed at 3-month intervals. Yellow, sickly leaves indicate deficiencies in minor elements, especially iron, in poor soils; apply minor element fertilizer to the planting bed at 3-month intervals.

DISADVANTAGES Established plantings may outgrow garden areas allotted to them.

Strelitziaceae
(Bird of Paradise Family)

The familiar orange-and-blue-flowered bird of paradise, **Strelitzia reginae,** is seen so often in Hawaiian floral decorations that it has very nearly assumed the status of a trademark for Hawaiian flowers. It is, in fact, a native of Africa, as are several others of the members of the bird of paradise family. Still other species are native to neighboring Madagascar and to tropical South America.

The birds of paradise formerly were classified as part of the **banana family (Musaceae),** as were the **heliconias.** The newly created family is quite small, consisting of three genera: *Ravenala* and *Strelitzia,* whose representative species are common in local gardens, and *Phenakospermum,* which is not known to have been introduced to Hawai'i. Of the eight species assigned to the family, three are described in this book.

The family's scientific name derives from the generic form *Strelitzia,* honoring Charlotte of Mecklenburg-Strelitz (1744–1818), German-born queen of George III of England.

Ravenala madagascariensis
Travelers' Tree

Many people think that the travelers' tree is a kind of palm because its many sturdy trunks very much resemble those of certain palms. This plant's leaves, however, tell us what it really is: they are distinctively banana-like in size and shape and help to show that it is a member of the bird of paradise family. The travelers' tree is a native of Madagascar, and now is cultivated widely throughout the world's tropics. The common name was given to it long ago because the plant accumulates drinkable water within its leaf bases. This water can be collected during times of drought or scarcity. The strong trunks and leaf midribs are used in constructing houses in parts of the Old World tropics. Dried leaves provide an adequate thatch. Young leaves and starchy fruits are eaten in Africa and Madagascar.

Young plants are very similar to those of the **giant birds of paradise, Strelitzia nicolai.** After many years of growth, however, they assume their characteristic tall, fan-topped appearance and the palmlike trunks. They are dramatic plants for tropical landscapes; single-trunked specimens provide some of the most spectacular silhouettes found in Hawaiian gardens; these specimens are produced by determined removal of any new root offshoots.

Ravenala is the plant's name in Madagascar; *madagascariensis* denotes its native habitat.

HABIT	A very large, erect, palmlike plant that produces many strong upright trunks, 2 feet in diameter, that can grow to nearly 75 feet in height. The plant's crown is tightly arranged in the shape of a gigantic flat fan made of several 10-foot banana-like leaves. Large green fan-shaped flower clusters appear within the leaf mass; individual flowers are about 3 feet in length; blue-fuzz-covered, edible seeds follow the flowers. Slow growth rate; easily transplanted.
GROWING CONDITIONS	Very adaptable, but will not withstand extreme salt conditions. Grows best in rich, well-watered, well-drained soil, in partial shade or full sunlight and protected from wind. Will withstand extreme drought conditions for considerable lengths of time.
USE	Specimen plant; tropical foliage and flowers; dry fruit clusters used in flower arrangements.
PROPAGATION	Generally by root offshoots, but may be started from seed.
INSECTS/DISEASES	To control oleander scale, use summer oil or malathion. Tall, established plantings are difficult to spray.
PRUNING	Remove unsightly dead and damaged leaves and flower clusters; plant is not naturally neat and tidy, drops its dead leaves reluctantly. Do not remove plant's crowns or damage its growing tips while pruning; removal of the crown will result in a dead stump. Clumps may be thinned by removing entire trunks.
FERTILIZING	Apply general garden fertilizer (10-30-10) to the planting bed at 4-month intervals.
DISADVANTAGES	Untrimmed dead leaves mar the plant's beauty. May be too large for most gardens.

Strelitzia nicolai
Giant Bird of Paradise, Blue and White Strelitzia,
Wild Banana

The giant bird of paradise is quite similar in appearance to the **travelers' tree** *(Ravenala madagascariensis)*. It has dense clusters of tall palmlike trunks that are topped with great fans of banana-like leaves. For this reason, in some tropical countries it is called wild banana. It is much smaller than the travelers' tree, however, growing only to about one-sixth the size of the larger plant. The giant bird of paradise is native to Natal and adjacent parts of the east coast of South Africa. This *Strelitzia* was introduced to European horticulture well over 100 years ago; the first specimen cultivated in Europe flowered in 1858. It was planted in the greenhouses of the Imperial Botanical Gardens, at St. Petersburg, Russia.

These are large plants for landscaping on a grand scale, producing as they do giant clumps of trunks and leaves, and giving a decidedly tropical jungle effect to any place in which they are grown. One single plant can develop into an enormous colony; older specimens may exhibit 20 or 30 trunks. The huge blue and white flowers, although interesting to look at, are not as attractive as are those of the common bird of paradise, **S. reginae,** nor are they used to such an extent in flower arrangements.

Strelitzia is named for Charlotte of Mecklenburg-Strelitz (1744–1818), German-born queen of King George III of England; *nicolai* honors Czar Nicholas I of Russia (1796–1855).

HABIT   A large, erect, palmlike plant that produces many strong trunks about 6 inches in diameter, each of which can be about 10 feet tall. The trunks are marked with distinctive, highly sculptured leaf scars; a fan of 6-foot banana-like leaves raises the plant's height to about 15 feet. Stemless blue and white flower clusters, resembling those of the bird of paradise, appear from gray and burgundy bracts; individual flowers are about 8 inches across; fruits follow the bloom. Slow growth rate; easily transplanted.

GROWING CONDITIONS   Adaptable, but will not withstand extreme salt conditions. Grows best in areas of rich, well-watered, well-drained soil, partial shade or full sunlight, and protected from winds. Will withstand a certain amount of drought.

USE   Specimen plant; mass planting; tropical flowers and foliage.

PROPAGATION   Generally from root offshoots, but may be started from seed.

INSECTS/DISEASES   To control oleander scale, use summer oil or malathion. Tall, established plants are somewhat difficult to spray.

PRUNING   Remove unsightly dead and damaged leaves and flower clusters; plant is not naturally neat and tidy, and sheds its dead leaves reluctantly. Do not remove plant's crowns or damage its growing tips while pruning; removal of the crown will result in a dead stump. Clump may be thinned by removing entire trunks.

FERTILIZING   Apply general garden fertilizer (10-30-10) to the planting bed at 4-month intervals.

DISADVANTAGES   Untrimmed dead leaves mar the plant's beauty.

170

Strelitzia reginae
Bird of Paradise, Crane's Bill, Crane Flower

Bird of paradise is quite different from its close relatives, the **giant bird of paradise *(Strelitzia nicolai),*** and **travelers' tree *(Ravenala madagascariensis).*** The two latter species produce treelike plants, whereas this one is a densely clumping shrub that grows to be little higher than a man's shoulders. It is interesting, moreover, in that it is ornithophilous, requiring nectar-eating birds to pollinate its blossoms, and is also protandrous, each flower's male and female parts being not concurrently receptive to each other. This fact of timing requires the nectar eaters to carry pollen from one flower's anthers to the receptive stigma of another before pollination can be accomplished.

The plant is native to the Cape of Good Hope region of South Africa, where often it grows wild along riverbanks. Although most admirers call it bird of paradise, some English-speaking people in Europe and Africa refer to it as crane's bill or crane flower. Whatever they may be, all vernacular names for it suggest the beaklike flower bracts and the flowers' resemblance to birds in flight.

The bird of paradise, one of our most popular plants, is grown either as a specimen or in masses in Hawai'i's sunnier garden locations. In addition, considerable commercial cultivation is done to supply the florist trade.

Strelitzia is named for Charlotte of Mecklenburg-Strelitz (1744–1818), German-born consort of George III of England; *reginae,* meaning of the queen, alludes to both Queen Charlotte and the regal plant.

HABIT A shrubby, clumping plant that grows to about 5 feet in height. Small, stiff, somewhat banana-like leaves sprout in dense clusters from short stems; individual leaves, 4 to 5 feet long, are blue-gray-green in color. Flowers appear during most of the year, especially in warmer months. Flower clusters are borne on sturdy stems; several orange and blue flowers are nestled in gray-green, beaklike bracts; individual new flowers appear as preceding ones age and wither. Orange-fuzz-covered seeds follow the flowers. Slow growth rate; easily transplanted.

GROWING CONDITIONS Very adaptable; will even grow near the beach if given some protection from strong salt winds. Grows best in rich, well-watered, well-drained soil, partial shade or full sunlight. Will withstand some drought and mistreatment.

USE Specimen plant; mass planting; tropical foliage and colorful flowers.

PROPAGATION Generally by division of the plant, but may be started from seed.

INSECTS/DISEASES To control oleander scale, use summer oil or malathion.

PRUNING Remove dead leaves and flower clusters; plant is rather neat in habit, requires little pruning; drastic pruning results in an unsightly mass that requires many months to recover.

FERTILIZING Apply general garden fertilizer (10-30-10) to the planting bed at 4-month intervals.

DISADVANTAGES Plant is often attacked by oleander scale.

Zingiberaceae
(Ginger Family)

The ginger family encompasses a large group of tropical herbs, most of which are native to the jungle forests and the fields of Indo-Malaysia. At present, the family consists of 45 genera and about 700 species. Chief among the genera are *Alpinia, Curcuma, Globba, Hedychium, Kaempferia, Nicolaia, Zingiber*—all of which are represented in this book—and *Amomum, Elettaria,* and *Renealmia.* Three species of great economic importance are commercial ginger *(Zingiber officinale)* which is cultivated for its fresh root, which is sold in Hawai'i's markets; cardamom *(Elettaria cardamomum),* whose seeds have been used as flavoring in India for millennia, and throughout much of the rest of the world for centuries; and turmeric *(Curcuma domestica),* one of the chief ingredients of curry powder and also the source of a widely used yellow dyestuff.

The edible gingers have been items of commerce since the dawn of history. Trade in gingers developed first throughout the Indonesian archipelago and along the coasts of the Indian Ocean. Later, gourmets in the Mediterranean area became acquainted with them. Early Greek and Roman writings recorded the use of some ginger spices; the Greek word *amomon* described one of the gingers of commerce at that early time. Today's international cuisine demands a constant supply of the flavorful herbs.

The family name Zingiberaceae is derived from *zingiberi,* a Greek word, used by Dioscorides, that apparently evolved from an ancient Indian name for ginger. Some Malays call it inchiver, from *inchi,* meaning root; this term probably stems from the same Indian word that Dioscorides borrowed. 'Awapuhi is the Hawaiian term for all the gingers.

Alpinia purpurata
Red Ginger, 'Awapuhi-'ula'ula

Red ginger is one of the most popular of tropical plants, both for landscaping and for cut flowers. Native to parts of Melanesia and the Moluccas, it is cultivated now in gardens throughout the tropical world. For festive occasions, South Pacific peoples use the colorful red flower bracts to make an ornamental dress, somewhat in the manner of chain mail. Bracts are also strung into leis. In Samoa, leis made of red ginger denote chiefly rank and are worn in important ceremonies.

Red ginger grows well in many locations, whether cool and shady or hot and sunny. Generally it is employed for tropical effect in informal jungle settings. In a process called epistasis, new plants will form along the stem of the matured flower stalk: the small young plants sprout roots as the flower fades, then drop to the ground to replenish the clump. One of the plant's great advantages is that it produces flowers throughout most of the year. Almost every commercial tropical flower arrangement includes two or three red gingers. The flowers are fairly good cut material; the older, hardened flower spikes last for a week or more.

Alpinia is named for Prosper Alpino (1553–1616), an Italian botanist; *purpurata* describes this plant's red-purple flower bracts. Red ginger in Hawaiian is 'awapuhi-'ula'ula.

HABIT   An upright, herbaceous, evergreen plant that may grow to about 12 feet in height. Several sturdy leaf stalks sprout from underground rhizomes; large, foot-long leaves are arranged along the upper stalks. Flowers appear mostly between spring and late fall; the terminal flower clusters consist of bright red bracts covering inconspicuous inch-long white flowers. Old stalks die back after flowering. New rooted plants sprout in numbers among the withered bracts of older flower stalks. Fast growth rate; easily transplanted.

GROWING CONDITIONS   Quite adaptable, but grows best in protected jungle-like environments in humus-rich, well-watered, well-drained soil; leaves should be somewhat protected from hot, drying winds; blooms well in areas of full sun or partial shade. Not a beach plant, but will grow near the sea if given utmost protection and constant moisture.

USE   Specimen plant; mass planting; tropical foliage and flowers; cut flower material.

PROPAGATION   By root division or by using rooted plantlets that develop among the faded flower clusters.

INSECTS/DISEASES   To control thrips, use diazinon or malathion.

PRUNING   Remove dead leaf and flower stalks after blossoming. New foliage grows back readily.

FERTILIZING   Apply general garden fertilizer (10-30-10) to the planting bed at 3-month intervals. Yellow, sickly leaves indicate deficiencies in minor elements, usually iron, in poor soils; apply minor element fertilizer to the planting bed at 3-month intervals.

DISADVANTAGES   Established plantings may outgrow garden areas allotted to them.

Alpinia purpurata cv. 'Jungle Queen'
Jungle Queen Ginger, 'Awapuhi

Jungle Queen, a cultivar of the common **red ginger *(Alpinia purpurata),*** is relatively new and rare in Hawaiian gardens. It is a native of Guadalcanal, Solomon Islands. Although the Jungle Queen and its parent are regarded as purely ornamental plants, several of their *Alpinia* relatives have long been involved in production of Oriental foods, fabrics, and medicines. The species most commonly used, *A. galanga,* provides an important spice, a dyestuff, and medicine in parts of Asia; it is always found in Chinese herb shops and was an item of early trade between Asia and Africa and the Mediterranean region. The root is commonly used as a food flavoring in Malaysia and neighboring regions; its dye, galangin, produces yellow and yellow-green colors in woolen fabrics when treated with the proper mordants. A tonic made from the root is used to relieve dyspepsia and similar stomach complaints.

Jungle Queen is grown in much the same way as is its parent. It, too, has developed the ability to reproduce through epistasis, the formation of new plants along the aging flower stalk.

Alpinia is named for Prosper Alpino (1553–1616), an Italian botanist; *purpurata* describes the parents' red-purple flower bracts. The common name, Jungle Queen, has been devised for the nursery and florist trades.

HABIT
: An upright, herbaceous, evergreen plant that may grow to about 12 feet in height. Several sturdy leaf stalks sprout from underground rhizomes; large, foot-long leaves are arranged along the upper stalks. Flowers appear mostly between spring and late fall; flower cluster consists of shell-pink bracts covering inconspicuous, inch-long white flowers. Old stalks die back after flowering. New rooted plants sprout in numbers among the withered bracts of older flower stalks. Fast growth rate; easily transplanted.

GROWING CONDITIONS
: Quite adaptable, but grows best in protected jungle-like environments in humus-rich, well-watered, well-drained soil; leaves should be somewhat protected from hot, drying winds; blooms well in partial shade. Not a beach plant, but will grow near the sea if given shade, humus soil, utmost protection, and considerable, constant moisture.

USE
: Specimen plant; mass planting; tropical foliage and flowers; excellent cut flower material.

PROPAGATION
: By root division or by using plantlets that develop at the flower clusters.

INSECTS/DISEASES
: To control thrips, use diazinon or malathion.

PRUNING
: Remove dead leaf and flower stalks after blossoming. New foliage grows back readily.

FERTILIZING
: Apply general garden fertilizer (10-30-10) to the planting bed at 3-month intervals. Yellow, sickly leaves indicate deficiencies in minor elements, usually iron, in poor soils; apply minor element fertilizer to the planting bed at 3-month intervals.

DISADVANTAGES
: Established plantings may outgrow garden areas allotted to them.

Alpinia sanderae
Sanders' Ginger, 'Awapuhi

This is one of the few kinds of gingers that are grown solely for their ornamental foliage. Its botanical genealogy is somewhat uncertain. It is a native of New Guinea, and may be a variegated-leaf form of *Alpinia rafflesiana,* a similar plant, native to Malaysia, which has leaves that are wholly green. *A. sanderae* is not known to have any practical use; however, its possible parent, *A. rafflesiana,* has leaves that are made into poultices for treating boils in Malacca. Two other species that have variegated green and white foliage are *A. vittatum,* another possible descendant of *A. rafflesiana* and a native of the tropical western Pacific; and *A. tricolor* from the Solomon Islands.

Sanders' ginger is among the smallest of the alpinias grown in Hawai'i; it rarely exceeds the height of a man. It is a colorful plant, even though it produces unremarkable flowers; often the bright green and white, unblemished foliage is used in the garden landscape to add brilliant spots of white to otherwise green backgrounds. It is excellent material for growing in overshaded locations, brightening dim areas where most flowering plants will not bloom.

Alpinia is named for Prosper Alpino (1553–1616), an Italian botanist; *sanderae* honors members of the Sanders family, nurserymen of St. Albans, England, and Bruges, Belgium.

HABIT An upright, herbaceous, evergreen plant that grows to about 8 feet in height; several sturdy leaf stalks sprout from underground rhizomes; large, 8-inch-long, variegated leaves are arranged openly along the leaf stalks. Old stalks die back after the flowering period. Fast growth rate; easily transplanted.

GROWING CONDITIONS Quite adaptable, but grows best in protected jungle-like environments, in humus-rich, well-watered, well-drained soil; leaves should be somewhat protected from hot, drying winds; blooms well in areas of full sun or partial shade. Not a beach plant, but will grow near the sea if given utmost protection and considerable, constant moisture.

USE Specimen plant; mass planting; tropical foliage and flowers; excellent cut flower material.

PROPAGATION By root division.

INSECTS/DISEASES To control thrips, use diazinon or malathion.

PRUNING Remove dead leaf and flower stalks after blossoming. New foliage grows back readily.

FERTILIZING Apply general garden fertilizer (10-30-10) to the planting bed at 3-month intervals. Yellow, sickly leaves indicate deficiencies in minor elements, usually iron, in poor soils; apply minor element fertilizer to the planting bed at 3-month intervals.

DISADVANTAGES Established plantings may outgrow garden areas allotted to them.

Alpinia zerumbet
Shell Ginger, Shell Flower, 'Awapuhi-luheluhe

Hawai'i's popular shell ginger is native to India and Burma. It is grown commonly throughout the tropics. In parts of Southeast Asia the stem pith of young shoots is eaten as a vegetable. Leaves of the plant are used as wrappers for foodstuffs, especially for cooked rice. During World War I, when flax was scarce, fibers from the shell ginger were employed as a substitute. Since then, little commercial use has been made of the fibers, but research has shown that they can be converted into papers of high quality. Shell ginger is not grown commercially in Hawai'i for the florist trade because other gingers provide better materials for cut flowers and leis. However, both leaves and flowers are used quite often in flower arrangements.

The plant is included in many jungle-like gardens; indeed, it has escaped cultivation in many places, and now grows in the cool, wet mountain forests of Hawai'i's largest islands. The vernacular name is most appropriate, for the individual rounded florets are similar to pink, yellow, and white sea shells.

Alpinia is named for Prosper Alpino (1553–1616), an Italian botanist; *zerumbet* is the Indian word for the wild species. Hawaiians call the plant 'awapuhi-luheluhe, drooping ginger; Southeast Asians know this and many other *Alpinia* species as languas, a Malay word.

HABIT   An upright, herbaceous, evergreen plant that grows to about 12 feet in height; several sturdy leaf stalks sprout from underground rhizomes; large, 2-foot leaves are arranged openly along the stalks. Flowers appear mostly from spring to late fall. The flowers are distinctively pendent and consist of many pink, shell-like corollas that enclose inch-long yellow and red flowers; round orange seed capsules follow the bloom. Old stalks die back after the flowering period. Fast growth rate; easily transplanted. A pure white flowered form is sometimes seen.

GROWING CONDITIONS   Quite adaptable, but grows best in protected jungle-like environments in humus-rich, well-watered, well-drained soil; leaves should be somewhat protected from hot, drying winds; blooms well in areas of full sun or partial shade. Not a beach plant, but will grow near the sea if given utmost protection and considerable, constant moisture.

USE   Specimen plant; mass planting; tropical foliage and flowers; excellent cut flower material.

PROPAGATION   By root division. New varieties may be produced from seed.

INSECTS/DISEASES   To control thrips, use diazinon or malathion.

PRUNING   Remove dead leaf and flower stalks after blossoming. New foliage grows back readily.

FERTILIZING   Apply general garden fertilizer (10-30-10) to the planting bed at 3-month intervals. Yellow, sickly leaves indicate deficiencies in minor elements, usually iron, in poor soils; apply minor element fertilizer to the planting bed at 3-month intervals.

DISADVANTAGES   Established plantings may outgrow garden areas allotted to them.

Curcuma roscoeana
Jewel of Burma, Ginger, 'Awapuhi

The jewel of Burma is a highly ornamental member of the economically important genus *Curcuma*. It is native to Indo-Malaysia, as are most of its related species.

Undoubtedly, the most important of the curcumas is turmeric, *C. domestica*, a Southeast Asian plant that has been cultivated for thousands of years throughout tropical Asia and the Pacific islands. Turmeric root is indispensable in the preparation of curries, being the most common of the many seasonings that enter into the innumerable kinds of combinations of spices and herbs that are called curry sauces. It is also one of the world's most important natural dyes, having long been used to give yellow color to both foods and fabrics. Early Polynesians brought turmeric—or 'olena—to Hawai'i. They used the root as a medicine, for dyeing tapa, and as a sacred component in religious rituals.

In appearance, jewel of Burma is quite like its more common counterpart, turmeric. The chief difference is that turmeric's flower bracts are nearly pure white, whereas in this plant the bracts are golden yellow at first, then turn to brilliant orange as they age. Both are tropical jungle plants.

Curcuma is derived from an Arabic word, *kurkum*, describing the yellow color found in the roots of many *Curcuma* species; *roscoeana* is named for William Roscoe (1753–1831), founder of the Liverpool Botanic Garden, England, in 1802, and author of *Monandrian Plants*, published in 1828.

HABIT A bushy, herbaceous, deciduous plant that grows to about 2 feet in height. Several short, sturdy leaf stalks sprout from underground rhizomes; foot-long leaves are closely arranged. Between late summer and early fall, a 5-inch flower cluster appears in the center of the leaf grouping; it consists of yellow-orange bracts enclosing inconspicuous yellow flowers. Leaf stalks die back completely after the flowering period. Fast growth rate; easily transplanted.

GROWING CONDITIONS Quite adaptable, but grows best in protected jungle-like environments in humus-rich, well-watered, well-drained soil; leaves should be somewhat protected from hot, drying winds; blooms well in full sun or partial shade. It is not a beach plant.

USE Specimen plant; mass planting; tropical foliage and flowers; cut flower material.

PROPAGATION By root division.

INSECTS/DISEASES To control thrips, use diazinon or malathion.

PRUNING Remove dead leaf and flower stalks after blossoming. New foliage grows back readily in the spring.

FERTILIZING Apply general garden fertilizer (10-30-10) to the planting bed at 3-month intervals. Yellow, sickly leaves indicate deficiencies in minor elements, usually iron, in poor soils; apply minor element fertilizer to the planting bed at 3-month intervals.

DISADVANTAGES Established plantings may outgrow garden areas allotted to them.

Globba atrosanguinea
Globba Ginger, 'Awapuhi

The *Globba* species are among the smallest of the ginger relatives; many are short, densely clumping plants that rarely grow to above 3 feet in height. The diminutive *G. atrosanguinea* is a native of Malaysia. It is one of about 50 species of the genus, all of which are endemic to parts of southern China and Indo-Malaysia. Several relatives have specific medicinal uses, primarily as aids to childbirth, and some are thought to have supernatural value in chasing away evil spirits. A relative, *G. marantina,* has edible roots that are used as seasoning in Malaysian cooking.

Several *Globba* species have been introduced to Hawai'i relatively recently; all are very similar to this plant, which has been grown ornamentally in the Islands for many years. The plants are tender forest-floor dwellers, requiring much protection from extreme sun and wind. They act very much like leafy groundcovers and may be used in handsome mass bedding arrangements. They are also excellent container plants. The small flower clusters are among the most spectacularly colored of all those produced by the gingers.

Globba is derived from the plant's native Amboinese name, galoba; *atrosanguinea,* from *ater,* meaning black or very dark, and *sanguinea,* meaning blood red, describes the burgundy flower stems and bracts.

HABIT
A small, bushy, herbaceous, deciduous plant that grows to about 3 feet in height. Many short, tender leaf stalks sprout in dense clumps from underground rhizomes; leaf stalks curve upward and outward in a fountain arrangement, each carrying several leaves 6 to 8 inches long. Blossoms appear mostly between spring and fall; unusual spidery, yellow flowers spring from stems borne on brilliant red-violet bracts. Old stalks die back after flowering. Lies dormant from November through March. Fast growth rate; easily transplanted.

GROWING CONDITIONS
Quite adaptable, but grows best in protected jungle-like environments in humus-rich, well-watered, well-drained soil; leaves should be somewhat protected from hot, drying winds; blooms well in areas of full sun or partial shade. Not a beach plant, but will grow near the sea if given utmost protection and considerable, constant moisture.

USE
Specimen plant; mass planting; tropical foliage and flowers; excellent cut flower material.

PROPAGATION
By root division.

INSECTS/DISEASES
To control thrips, use diazinon or malathion.

PRUNING
Remove dead leaf and flower stalks after blossoming. New foliage grows back readily.

FERTILIZING
Apply general garden fertilizer (10-30-10) to the planting bed at 3-month intervals. Yellow, sickly leaves indicate deficiencies in minor elements, usually iron, in poor soils; apply minor element fertilizer to the planting bed at 3-month intervals.

DISADVANTAGES
Plant is dormant during part of the year.

Hedychium flavescens
Yellow Ginger, Gandasuli, 'Awapuhi-melemele

The fragrant yellow ginger comes from Bengal State, India; it and its Indo-Chinese twin, white ginger, *Hedychium coronarium*, have been cultivated in tropical Asia and India since ancient times. The two plants are universally loved for the exquisite fragrance of their flowers. The yellow ginger's Sanskrit name is gandasuli, or fragrance of the princess. At one time fibers from leaves and stems were made into fine paper in the Tonkin region; but the lack of adequate amounts of plant material makes this process commercially unfeasible today. Moluccans sometimes use parts of the plant to reduce swellings and assuage sore throats.

Yellow ginger ('awapuhi-melemele in Hawaiian) and white ginger ('awapuhi-ke'oke'o) are among the most favorite of flowers for making leis. Their fragrance is pervasive: just a few will perfume an entire room. Both species have escaped from cultivation and grow wild in the wet forest regions of Hawai'i's major islands. Such lush invasions of the landscape are so ubiquitous and so extensive that many residents believe these gingers have been growing in the Islands since the beginning of time. Actually, they were imported to Hawai'i late in the nineteenth century.

Hedychium, from *hedys*, meaning sweet, and *chion*, meaning snow, actually describes the white ginger's flowers; *flavescens* means pale yellow; *coronarium*, meaning crown, refers to the coronet of flowers within the cluster.

HABIT
: An upright, herbaceous, evergreen plant that grows to about 7 feet in height. Several sturdy leaf stalks sprout from underground rhizomes; leaves, nearly 2 feet long, are arranged loosely along the stalks. Flowers appear mostly from late spring to late fall; individual 3-inch flowers, light yellow and very fragrant, stand out from the green bracts of the central cluster. Old stalks die back after the flowering period. Fast growth rate; easily transplanted.

GROWING CONDITIONS
: Quite adaptable, but grows best in protected jungle-like environments in humus-rich, well-watered, well-drained soil; leaves should be somewhat protected from hot, drying winds; blooms well in full sun or partial shade. It is not a beach plant.

USE
: Specimen plant; mass planting; tropical foliage and flowers; excellent cut flower material.

PROPAGATION
: By root division.

INSECTS/DISEASES
: To control thrips, use diazinon or malathion.

PRUNING
: Remove dead leaf and flower stalks after blossoming; generally the entire clump is cut to the ground after the flowering period. New foliage grows back readily.

FERTILIZING
: Apply general garden fertilizer (10-30-10) to the planting bed at 3-month intervals. Yellow, sickly leaves indicate deficiencies in minor elements, usually iron, in poor soils; apply minor element fertilizer to the planting bed at 3-month intervals.

DISADVANTAGES
: Established plantings may outgrow garden areas allotted to them.

Hedychium gardnerianum
Kahili Ginger, 'Awapuhi

Hedychium gardnerianum has an interesting botanical history. It is native to Nepal and Sikkim, which in the nineteenth century were closed and forbidden kingdoms. In the early 1800s Dr. Nathaniel Wallich, noted Danish explorer and plant collector, through the good offices of the East India Company's resident in Nepal, was able to organize collecting parties into the Himalayas. Wallich found this plant growing in the valley of Katmandu. Subsequently, Sir Joseph D. Hooker, eminent botanist and director of the Royal Botanic Gardens at Kew in England, found plantings in Sikkim. Specimens of those discoveries were grown initially in the botanic gardens at Calcutta, and in 1823 were introduced to England.

The plant is commonly known in Hawai'i as kahili ginger. Its open, cylindrical flower clusters resemble in form the great royal insignia of Hawai'i, made of feathers from native birds, and set on long poles—the *kāhilis* that preceded and flanked the ruling chiefs during ceremonial occasions. A number of flower variations can be found. The plant has escaped from cultivation and become naturalized in the Volcano area of the Big Island.

Hedychium, from *hedys*, meaning sweet, and *chion*, meaning snow, describes the white ginger's flowers; *gardnerianum* honors Colonel Edward Gardner, the East India Company's resident who was able to facilitate Wallich's entry into Nepal.

HABIT An upright, herbaceous, evergreen plant that grows to about 8 feet in height. Several sturdy leaf stalks sprout from underground rhizomes; foot-long leaves grow in open arrangement along the stalks. Blossoms appear mostly from late spring to late fall; individual 3-inch flowers, yellow and fragrant, have brilliant red, long stamens which give the entire flower cluster a burnt orange cast. Flowers spring out from the green bracts of the central cluster in a large, foot-high, cylindrical form. Old stalks die back after the flowering period. Fast growth rate; easily transplanted.

GROWING CONDITIONS Quite adaptable, but grows best in protected jungle-like environments in humus-rich, well-watered, well-drained soil; leaves should be somewhat protected from hot, drying winds; blooms well in full sun or partial shade. It is not a beach plant.

USE Specimen plant; mass planting; tropical foliage and flowers; cut flower material.

PROPAGATION By root division.

INSECTS/DISEASES To control thrips, use diazinon or malathion.

PRUNING Remove dead leaf and flower stalks after blossoming; generally the entire clump is cut to the ground after the flowering period. New foliage grows back readily.

FERTILIZING Apply general garden fertilizer (10-30-10) to the planting bed at 3-month intervals. Yellow, sickly leaves indicate deficiencies in minor elements, usually iron, in poor soils; apply minor element fertilizer to the planting bed at 3-month intervals.

DISADVANTAGES Established plantings may outgrow garden areas allotted to them.

Hedychium greenei
Greene's Ginger, 'Awapuhi

Hedychium greenei is much like its somewhat larger relative, the **kahili ginger (*H. gardnerianum)*;** indeed, it is difficult to distinguish between the two plants. Both are natives of India, and both have the familiar kahili flower-cluster form. The primary distinguishing characteristic is in the flower color: whereas kahili ginger produces brilliant yellow flowers with bright red central filaments, the flowers of Greene's ginger are of a more subdued apricot hue. These hedychiums and others like them generally are grown for their ornamental fragrant blossoms. Some species produce fibers suitable for making fine writing papers, and one, *H. longicornutum,* has roots that in Indo-Malaysia are processed into a remedy for earache.

This plant, like most of the larger ornamental gingers, grows well in protected, jungle-like environments. It is used—rather less often than are its *Hedychium* relatives—in massive background plantings where the bright, dense foliage can provide visual screening, and where the flowers can be enjoyed during the long blooming season.

Hedychium, from *hedys,* meaning sweet, and *chion,* meaning snow, most nearly describes the white ginger (*H. coronarium,* p.190); *greenei* is named for a Dr. Greene (1793–1862) of Boston.

HABIT An upright, herbaceous, evergreen plant that grows to about 6 feet in height. Several sturdy leaf stalks sprout from underground rhizomes; leaves, each about 10 inches long, are arranged in open fashion along the stalks. Flowers appear mostly from late spring to late fall; individual 2-inch, fragrant, orange flowers are arranged densely on stems rising from green bracts in a foot-high cylindrical form. Old stalks die back after the flowering period. Fast growth rate; easily transplanted.

GROWING CONDITIONS Quite adaptable, but grows best in protected jungle-like environments in humus-rich, well-watered, well-drained soil; leaves should be somewhat protected from hot, drying winds; blooms well in areas of full sun or partial shade. Not a beach plant, but will grow near the sea if given utmost protection and considerable, constant moisture and humus soil.

USE Specimen plant; mass planting; tropical foliage and flowers; excellent cut flower material.

PROPAGATION By root division.

INSECTS/DISEASES To control thrips, use diazinon or malathion.

PRUNING Remove dead leaf and flower stalks after blossoming; generally the entire clump is cut to the ground after the flowering period. New foliage grows back readily.

FERTILIZING Apply general garden fertilizer (10-30-10) to the planting bed at 3-month intervals. Yellow, sickly leaves indicate deficiencies in minor elements, usually iron, in poor soils; apply minor element fertilizer to the planting bed at 3-month intervals.

DISADVANTAGES Established plantings may outgrow garden areas allotted to them.

Kaempferia ovalifolia
Ginger, 'Awapuhi

The natural range of plants in the genus *Kaempferia* is quite wide, extending from tropical Africa to India, Malaysia, and southern China. About 70 species are recognized in this rather large plant group. *K. ovalifolia* is native to Thailand, Malaysia, and Malacca, and grows in profusion along the Tenasserim River in Burma. Several *Kaempferia* species have practical uses. The tubers of *K. aethiopica,* from Africa, are employed in seasoning foods, just as commercial ginger *(Zingiber officinale)* is used in countries farther to the east; and an Indian relative, *K. galanga,* or false galanga, is used throughout Indo-Malaysia to flavor rice. Several species yield medicinal preparations, especially for the healing or cicatrizing of wounds and the poulticing of boils and other skin ailments.

This plant is the shy cousin of the many gingers grown in Hawaiian gardens. It is extremely tender and must be grown in secluded, protected garden areas where heavy sun, wind, or debris from trees cannot harm it. It is a short, clumping groundcover, that grows profusely during the warmer months, then dies back to the ground for a long period during the winter.

Kaempferia is named for Engelbert Kaempfer (1651–1716), a German physician who traveled to parts of China and Japan, and who wrote about life in Japan and described many of its plants; *ovalifolia* means oval leaves.

HABIT	A small, bushy, herbaceous, deciduous plant that grows to about 14 inches in height. Several short, tender leaf stalks sprout from underground rhizomes; rounded, banded, dark green leaves, each about 8 inches long, are compactly arranged in a dense mass. Blossoms appear mostly from late spring to early fall; white and lavender flowers, each 2 inches in diameter, spring from checkered bracts hidden in the foliage. The crown dies back to the ground in late winter, revives in the spring. Fast growth rate; easily transplanted.
GROWING CONDITIONS	Extremely tender; grows best in protected jungle-like environments in humus-rich, well-watered, well-drained soil; leaves should be well protected from hot, drying winds; blooms well in areas of full or partial shade.
USE	Specimen plant; mass planting; tropical foliage and flowers; excellent cut flower material.
PROPAGATION	By root division.
INSECTS/DISEASES	To control thrips, use diazinon or malathion.
PRUNING	Remove dead leaf and flower stalks in the fall after blossoming, when leaves begin to wither. New foliage grows back readily in the spring.
FERTILIZING	Apply general garden fertilizer (10-30-10) to the planting bed at 3-month intervals. Yellow, sickly leaves indicate deficiencies in minor elements, usually iron, in poor soils; apply minor element fertilizer to the planting bed at 3-month intervals.
DISADVANTAGES	Dormant during winter months.

Nicolaia elatior
Torch Ginger, 'Awapuhi-ko'oko'o

The spectacular torch ginger is native to the jungles of Malaysia and Indonesia. It is one of the most strikingly beautiful of all tropical flowering plants. Two forms are planted throughout the world, the red-flowered one shown here and one with pink blossoms. Several species of *Nicolaia*, including this one, are cultivated in Malaysia for the seasoning of foods; their young flower shoots are used also as an ingredient in preparing curries. Fibers from the leaves and stems of another relative, *N. hemisphaerica*, could be processed into a useable paper; and *N. heyneana*, having properties that are considered to be tonic in Java, is commonly used as a medicine for ailing water buffaloes.

One of the interesting characteristics of the *Nicolaia* species is the fact that their flower stalks rise directly out of the ground, being completely separated from the leaf stalks. The torch ginger's brilliant cerise flower stalks sometimes grow to eye level, terminating in great waxy heads that, on first inspection, look utterly artificial.

Nicolaia is named for Czar Nicholas I of Russia (1796–1855); *elatior*, meaning taller, describes the plant's erect habit. 'Awapuhi-ko'oko'o means walking-stick ginger.

HABIT A large upright, herbaceous, evergreen plant that grows to about 20 feet in height. Leaves, about 2-feet long, are arranged evenly along tops of the tall stalks. Blossoms appear mostly from late spring to late fall, but occasionally in winter; tall torchlike flowers are red, pink, or white, and are borne on 5-foot stems that sprout directly from the root system. Flowers can be 8 inches in diameter. Older stalks die periodically; new growth appears constantly. Fast growth rate; easily transplanted.

GROWING CONDITIONS Quite adaptable, but grows best in protected environments in humus-rich, well-watered, well-drained soil; leaves should be somewhat protected from hot, drying winds; blooms well in areas of full sun or partial shade. Not a beach plant, but will grow near the sea if given utmost protection and considerable, constant moisture.

USE Specimen plant; mass planting; tropical foliage and flowers; excellent cut flower material.

PROPAGATION By root division.

INSECTS/DISEASES To control thrips, use diazinon or malathion.

PRUNING Remove dead leaf and flower stalks after blooming.

FERTILIZING Apply general garden fertilizer (10-30-10) to the planting bed at 3-month intervals. Yellow, sickly leaves indicate deficiencies in minor elements, usually iron, in poor soils; apply minor element fertilizer to the planting bed at 3-month intervals.

DISADVANTAGES Established plantings may outgrow garden areas allotted to them.

Nicolaia hieroglyphica
Ginger, 'Awapuhi

The genus *Nicolaia* consists of a small group of gingers native to the Indo-Malaysian region. At present, 25 species are recognized. Formerly, the genus was called *Phaeomeria*, a name that may be more familiar to many gardeners. The most popular of the ornamental *Nicolaia*s is the **torch ginger (Nicolaia elatior).** Several species are useful in tropical Asia, especially *N. speciosa*, which is widely cultivated for its edible flower and fruit shoots and seeds. Preparations of these plant parts often are included in curries and condiments. They possess a tart flavor somewhat similar to that of tamarind fruits *(Tamarindus indicus)* that give an acid pungency to Oriental foods. Some parts of *Nicolaia speciosa* are used medicinally also: the fruit is processed into a remedy for earache, and the leaves have purging properties. Stems are plaited into matting in Sumatra.

N. hieroglyphica is one of the larger gingers: its great leaf stalks arch above a man's head. But the flowers are somewhat small by *Nicolaia* standards, growing only a foot or two above the ground, and look like little clubs. Considerable garden area is required for established plantings, and the tall leaves must be protected from heavy, drying winds.

Nicolaia is named for Czar Nicholas I of Russia (1796–1855); *hieroglyphica* describes the flower markings that suggest ancient Egyptian writing.

HABIT
A tall, upright, herbaceous, evergreen plant that grows to about 15 feet. Two-foot-long leaves are arranged evenly along the tops of the tall stalks; leaves are bright green on upper surfaces and burgundy-red underneath. Flowers appear mostly from late spring to early fall; short-stemmed, dark red and yellow flower clusters sprout directly from the root system; clublike flower clusters reach about 3 inches in diameter. Many fruits follow the blossoms. Older flower and leaf stalks die back periodically; new growth appears constantly. Fast growth rate, easily transplanted.

GROWING CONDITIONS
Quite adaptable, but grows best in protected environments in humus-rich, well-watered, well-drained soil.

USE
Specimen plant; mass planting; tropical foliage and flowers; excellent cut flower material.

PROPAGATION
By root division or from seed.

INSECTS/DISEASES
To control thrips, use diazinon or malathion. Plant sometimes is subject to attack by Chinese rose beetles; no practical method of control is effective for a plant of this size.

PRUNING
Remove dead leaf and flower stalks after blossoming. New foliage grows back readily.

FERTILIZING
Apply general garden fertilizer (10-30-10) to the planting bed at 3-month intervals. Yellow, sickly leaves indicate deficiencies in minor elements, usually iron, in poor soils; apply minor element fertilizer to the planting bed at 3-month intervals.

DISADVANTAGES
Established plantings may outgrow garden areas allotted to them.

Zingiber spectabile
Ginger, Nodding Gingerwort, 'Awapuhi

This lovely plant is seen rather rarely in Hawaiian gardens, although it is one of the most ubiquitous of the 80 or so *Zingiber* species found in nature. The forests of its native Malaysia abound with the plant. As with a better known relative, commercial ginger *(Z. officinale)*, parts of this plant too are used as flavoring in Malay cooking. It is also respected for its curative powers: the pounded leaves act as an efficient poulticing agent and, when steeped in cold water, produce a soothing eyewash.

The plant is aggressive in a jungle habitat, and when provided these conditions in a garden, it will fill extensive areas if allowed to grow unchecked. It has not yet escaped from cultivation in Hawai'i, but more than likely our lush valleys will be invaded before long. The Malays know the plant as tĕpus tundok, nodding gingerwort, because the supple leaves and leaf stalks do nod gently in air currents. The bright yellow flower heads are beautifully formed, seeming to be carved from beeswax. They provide excellent material for flower arrangements, but so far are not grown in sufficient number to supply the florist trade.

Zingiber is a Greek name used by Dioscorides to describe a ginger with which he was familiar; the word stems from an ancient Indian name for gingers; *spectabile,* meaning admirable, describes the flower heads.

HABIT — A sturdy, upright, herbaceous, evergreen plant that grows to about 6 feet in height. Dark green leaves, each about 1 foot long, are arranged evenly at tops of tall stalks. Flowers appear mostly from late spring to early fall; 8-inch flower clusters appear on 2-foot stems and are composed of curling, waxy yellow bracts in which nestle small yellow and white flowers. Old stalks die back after the blooming period. Fast growth rate; easily transplanted.

GROWING CONDITIONS — Quite adaptable, but grows best in protected jungle-like environments in humus-rich, well-watered, well-drained soil; leaves should be somewhat protected from hot, drying winds; blooms well in areas of partial shade. Not a beach plant, but will grow near the sea if given utmost protection and considerable, constant moisture and humus soil.

USE — Specimen plant; mass planting; tropical foliage and flowers; excellent cut flower material.

PROPAGATION — By root division.

INSECTS/DISEASES — To control thrips, use diazinon or malathion.

PRUNING — Remove dead leaf and flower stalks after blossoming. New foliage grows back readily.

FERTILIZING — Apply general garden fertilizer (10-30-10) to the planting bed at 3-month intervals. Yellow, sickly leaves indicate deficiencies in minor elements, usually iron, in poor soils; apply minor element fertilizer to the planting bed at 3-month intervals.

DISADVANTAGES — Established plantings may outgrow garden areas allotted to them.

200

Zingiber zerumbet cv. 'Darceyi'
Variegated Wild Ginger

The variegated wild ginger is a cultivar of a widely grown and much-used plant, *Zingiber zerumbet*. The parent, native to the vast region that stretches from the Himalayas southward to Sri Lanka and eastward through the Malay-Indonesian archipelago, has been distributed to other tropical regions by migrating people. Hawai'i's Polynesian settlers brought the green-leafed plant to these islands centuries ago. Here, it has long since escaped from cultivation and may be found growing wild in valleys and on forested slopes. Hawaiians used the foamy flower heads for shampooing the hair and quenching thirst. Javanese use the rhizomes as a condiment for food and as a medicine for coughs and asthma; it is commonly seen in Indonesian markets.

Both the parent and its variegated cultivar are low plants, as gingers go. They send up a small cluster of hip-high leaves during the warm spring and summer months; the small pinecone-shaped flower heads appear on short stems during the late fall. The presence of the cleansing foam is easily established on the surface of the ripe flower clusters by gentle rubbing. The shampoo fluid is lightly scented with ginger.

Zingiber is a Greek name used by Dioscorides to describe a ginger with which he was familiar; the Greek word stems from an earlier Indian name for the plant group; *zerumbet* is the Indian word for the parent.

HABIT
A small upright, herbaceous, deciduous plant that grows to about 3 feet in height. Several rather tender leaf stalks sprout in a dense cluster from underground rhizomes; variegated green and white leaves, each about 6 inches long, are arranged evenly along the stalks. Flowers appear in the fall; compact, cylindrical flower clusters about 3 inches long, are composed of bracts that are initially green and become bright red, and from which issue papery white, orange, or yellow flowers. Plant dies back during winter. Fast growth rate; easily transplanted.

GROWING CONDITIONS
Quite adaptable, but grows best in protected environments in humus-rich, well-watered, well-drained soils; leaves should be somewhat protected from hot, drying winds; blooms well in areas of full sun or partial shade. Not a beach plant, but will grow near the sea if given utmost protection and considerable, constant moisture.

USE
Specimen plant; mass planting; tropical foliage and flowers; excellent cut flower material.

PROPAGATION
By root division.

INSECTS/DISEASES
To control thrips, use diazinon or malathion.

PRUNING
Remove dead leaf and flower stalks after blossoming. New foliage grows back readily.

FERTILIZING
Apply general garden fertilizer (10-30-10) to the planting bed at 3-month intervals. Yellow, sickly leaves indicate deficiencies in minor elements, usually iron, in poor soils; apply minor element fertilizer to the planting bed at 3-month intervals.

DISADVANTAGES
Established plantings may outgrow garden areas allotted to them.

Costaceae
(Costus Family)

The taxonomists, through the centuries, have succeeded in thoroughly confounding the nomenclature of this group of noble plants. The plant with the earliest known claim to the name *Costus* is a member of the daisy family (Compositae) now named *Saussurea lappa*, and, botanically speaking, is far removed from the plants in the present-day genus *Costus*. *Saussurea lappa*, known to Pliny and Dioscorides as *Costus arabicus* (even though it is native to Kashmir), was, and is, the source of a primary component in Oriental perfumes and incenses. Later confusion occurred when a relative of the gingers was misidentified as being the ''Arabian costus,'' whereupon Europe's botanists appropriated the old name for this entirely new group of plants. The *Costus* group was long considered to be an integral part of the **ginger family (Zingiberaceae),** but now these plants are established in a family of their own.

Four genera and more than 200 species make up the family. Its representatives are found throughout much of the tropical world, from Africa through Indo-Malaysia to Central and South America. In tropical Africa the ginger lily *(C. afer)* provides material for baskets and medicines for the relief of nausea. *C. lucanusianus* contains substances that congeal the latex sap of *Landolphia owariensis,* a large African periwinkle that is the source of the rubber called *rouge du Congo.* The tropical American *Costus spicatus* has sap that is taken medicinally as a diuretic. Probably the most important member of the family is the **crepe ginger *(C. speciosus).*** Three other genera complete the family: *Dimerocostus, Monocostus,* and *Tapeinochilus.*

Costus speciosus
Crepe Ginger, Setawar, Malay Ginger, 'Awapuhi

Costus speciosus was thought by the seventeenth-century Dutch botanist Kasper Commelin to have been the plant that the sixteenth-century Italian botanist Pierandrea Mattioli misidentified as *C. arabicus,* thereby setting the stage for future botanical confusion (see p. 207). This plant, commonly known throughout much of the English-speaking world as crepe ginger, has been used for centuries in its native Indo-Malaysia as a remedy for numerous afflictions and maladies including colds, fevers, pneumonias, and rheumatism. The rhizomes are edible and nutritious, although less palatable than some other kinds of ginger roots, and for this reason are reserved for times of famine. The young shoots, when cooked with coconut milk, are eaten much more often as a vegetable. The Malays, who call it setawar, believe it has supernatural powers as a potent antidote to evil spirits.

Crepe ginger is a popular garden plant in Hawai'i; its brilliant red, cone-like flower head produces the delicate and frilly white flowers which give the plant its common name. Like most of its near relatives among the gingers, it enjoys lush, protected, jungle-like environments. The flower heads generally are not used for arrangements; therefore, it is not grown commercially.

Costus is an old Greek name for an unrelated plant known to Pliny and Dioscorides; *speciosus,* meaning showy, describes the flowers.

HABIT
: An upright, herbaceous, evergreen plant that grows to about 6 feet in height. Several very sturdy leaf stalks sprout from underground rhizomes; leaves, each about 8 inches long, are grouped together at the top of the stalks in a distinctive spiral arrangement. Flowers appear mostly from late spring to early fall and are produced within the crown of leaves; small green bracts form a tight base for white and red-orange ruffled blossoms, each about 2 inches in diameter. Old flower stalks die back after the blooming period. Fast growth rate; easily transplanted.

GROWING CONDITIONS
: Quite adaptable, but grows best in protected environments, in humus-rich, well-watered, well-drained soil; leaves should be somewhat protected from hot, drying winds; blooms well in areas of full sun or partial shade. Not a beach plant, but will grow near the sea if given utmost protection and considerable, constant moisture, and humus soil.

USE
: Specimen plant; mass planting; tropical foliage and flowers.

PROPAGATION
: By root division.

INSECTS/DISEASES
: To control thrips, use diazinon or malathion.

PRUNING
: Remove dead leaf and flower stalks after blooming. New foliage grows back readily.

FERTILIZING
: Apply general garden fertilizer (10-30-10) to the planting bed at 3-month intervals. Yellow, sickly leaves indicate deficiencies in minor elements, usually iron, in poor soils; apply minor element fertilizer to the planting bed at 3-month intervals.

DISADVANTAGES
: Established plantings may outgrow garden areas allotted to them.

Costus spicatus
Indianhead Ginger, Spiral Flag

The Indianhead ginger is native to the islands of the Caribbean, southern Mexico, and tropical South America. It is used by people of those regions as a medicine: its sap is said to be an effective diuretic. The plant somewhat resembles the more familiar **crepe ginger** *(Costus speciosus),* but is much larger and more attenuated. As is typical of many *Costus* relatives, the leaves of this one too are arranged spirally along the twisted stems. For this reason, some people call it spiral flag.

It is a forest plant and grows happily in deeply shaded jungle environments, often attaining a height well above a man's head, where much of the foliage appears. Established plantings are extremely dense and all but impenetrable, being made up of hundreds of bare, canelike stems. It is best used as a background, where the attractive, glossy foliage and brilliant flower heads with their red bracts may be seen easily. The bright flowers stand out dramatically against the otherwise green landscape.

Costus is an old Greek name for an unrelated plant known to Pliny and Dioscorides (about the first century A.D.); *spicatus,* meaning spikelike, refers to the stiff, erect flower cluster.

HABIT An upright, herbaceous, evergreen plant that grows to about 10 feet in height. Several very sturdy leaf stalks sprout from underground rhizomes; glossy leaves, each about 8 inches long, are grouped at the tops of stalks in a distinctive spiral arrangement along the stems. Flowers appear at stem tips mostly from late spring to early fall, and sporadically throughout much of the rest of the year; large, brilliant red bracts form a tight base for small yellow and red flowers. Old flower stalks die back after the blooming period. Fast growth rate; easily transplanted.

GROWING CONDITIONS Quite adaptable; but grows best in protected environments in humus-rich, well-watered, well-drained soil; leaves should be somewhat protected from hot, drying winds; blooms well in areas of full sun or partial shade. Not a beach plant, but will grow near the sea if given utmost protection and considerable, constant moisture.

USE Specimen plant; mass planting; tropical foliage and flowers; excellent cut flower material.

PROPAGATION By root division.

INSECTS/DISEASES To control thrips, use diazinon or malathion.

PRUNING Remove dead leaf and flower stalks after blossoming. New foliage grows back readily.

FERTILIZING Apply general garden fertilizer (10-30-10) to the planting bed at 3-month intervals. Yellow, sickly leaves indicate deficiencies in minor elements, usually iron, in poor soils; apply minor element fertilizer to the planting bed at 3-month intervals.

DISADVANTAGES Established plantings may outgrow garden areas allotted to them.

Tapeinochilus ananassae
Indonesian Ginger

The so-called Indonesian ginger is native to the Moluccas, and, of course, is not a ginger but a costus. About 20 species are assigned to the genus *Tapeinochilus,* all of which are endemic to one part or another of Indonesia, New Guinea, the Bismarck Archipelago, and Queensland, Australia. Often the colorful flower heads are sold in Moluccan markets. This species was introduced to Hawai'i in 1959 by Joseph F. Rock, and has since enjoyed an ever-growing popularity as fine florist material.

The flower heads verge on the unreal, seeming to be made of plastic. This same effect is shown by several other gingers and costuses, and also by some of the anthuriums. This plant's brilliant red, waxy, unyielding, thorn-tipped bracts ward off most predators, both insect and herbivorous. Because this is a plant that evolved in the hot, humid Moluccan jungles, in Hawai'i it grows best in extremely moist locations, especially those protected by forests. Ideal areas in the Islands are at Hāna, Maui, and near Hilo, Hawai'i.

Tapeinochilus, from *tapeinos,* meaning low, and *cheilos,* meaning lip, refers to the small-lipped and quite insignificant yellow flowers that hide deep within the protecting red bracts; *ananassae,* meaning *Ananas*-like, refers to the resemblance of the flower heads to the fruit of a pineapple.

HABIT
A tender-leafed, herbaceous, evergreen plant that grows to about 8 feet in height. Several leaf stalks sprout from underground rhizomes at considerable distance from each other; leaves about 18 inches long are loosely arranged along the stalks. Blossoms appear mostly from late spring to early fall; 3- to 4-foot flower stems sprout directly from the root system; flower clusters 8 inches long are composed of stiff, waxy, spiny, red bracts in which nestle tiny yellow flowers. Sometimes flower clusters are topped by a small leafy stem. Old stalks die back during the winter months. Fast growth rate; easily transplanted.

GROWING CONDITIONS
Grows best in extremely moist, protected jungle environments in humus-rich, well-watered, well-drained soil; leaves should be protected from hot, drying winds; blooms well in areas of partial shade.

USE
Specimen plant; mass planting; tropical foliage and flowers; excellent cut flower material.

PROPAGATION
By root division, or from offshoots along woody stems.

INSECTS/DISEASES
To control thrips, use diazinon or malathion.

PRUNING
Remove dead leaf and flower stalks after blossoming. New foliage and flower shoots sprout readily.

FERTILIZING
Apply general garden fertilizer (10-30-10) to the planting bed at 3-month intervals. Yellow, sickly leaves indicate deficiencies in minor elements, usually iron, in poor soils; apply minor element fertilizer to the planting bed at 3-month intervals.

DISADVANTAGES
Established plantings may outgrow garden areas allotted to them.

Cannaceae
(Canna Family)

The canna family, pantropic in origin, consists of the single genus *Canna* and about 55 species. Several are ornamental. The species most important to agriculture is *C. edulis,* a native of Central and South America. Because its roots are rich in starch, this plant is cultivated commercially in Australia for the product called Queensland arrowroot, and in the Caribbean for a similar thickening agent known as *tous-les-mois.* The plant's tubers are quite similar in taste to white potatoes when cooked, although they are not as palatable because of their high fiber content. Also, the green leaves and stems make good cattle feed. Other valued species include **C. indica;** *C. bidentata,* whose seeds are made into necklaces and whose leaves are used as food wrappings in tropical Africa; *C. glauca* and *C. gigantea,* two Brazilian species, the first of which is used as a cooked vegetable, the second as a diuretic medicine; and *C. speciosa,* which is cultivated in Sierra Leone as the African equivalent of turmeric.

The family name Cannaceae is derived from that of the genus, *Canna,* the etymology of which is uncertain. The Latin word *canna* means reed; perhaps because the cannas' vertical leaf stems are reedlike in appearance, the Latin word was chosen to identify them. Another explanation might lie in the fact that Buddhists in India sometimes wear rosaries made from canna seeds. To them, *canna* means help from Buddha.

Canna indica
Canna, Indian Shot, Ali'ipoe, Li'ipoe

This popular garden plant is a product of tropical America and the Caribbean. Some confusion surrounds its origin because its scientific name, *Canna indica*, suggests that the plant comes from India, rather than from the West Indies.

This canna was introduced early into other parts of the tropics. It has escaped from cultivation and grows wild in many places. The ripe seeds are strung into rosaries used by both Buddhists and Mohammedans. Hawaiians, who make ornamental leis from the seeds, know the plant as ali'ipoe or li'ipoe, both meaning tiny globes. A number of these seeds are inserted into a hollow fruit of the calabash tree *(Crescentia cujete)* to make a gourd-rattle, or *'uli'uli,* used in certain hula dances.

In the garden, cannas generally are planted in masses in places receiving full sunlight. Flower and leaf variations abound; many have been given varietal names.

Canna derives either from the Latin word for reed, or from an Indian word denoting help from Buddha; *indica* means, in this case, native to tropical America, the Indies of Columbus.

HABIT
: A tender, herbaceous, partly deciduous plant that grows to about 5 feet in height; tender, vertical, fleshy stems support several light green leaves, each about 18 inches long, and a dense cluster of large, frilly, lily-like blossoms; flower colors range from white through deep yellow to pinks and reds; seeds follow in quantity. Fast growth rate; easily transplanted.

GROWING CONDITIONS
: Highly adaptable; will grow almost anywhere in Hawai'i, even at the beach if given protection from strong salt winds. Grows best in rich, well-watered, well-drained soil, in full sun or slightly shaded locations.

USE
: Specimen plant; mass planting; colorful flowers.

PROPAGATION
: Plants of a particular flower color are propagated by root division; new flower forms are developed from seed.

INSECTS/DISEASES
: To control thrips and mealybugs, use diazinon or malathion. For rust disease on foliage in wet areas, remove affected leaves to prevent spread and apply captan to healthy foliage.

PRUNING
: Remove dead leaf and flower stalks after blossoming; generally the entire clump is cut to the ground after the flowering period is completed. New foliage grows back readily in the spring.

FERTILIZING
: Apply general garden fertilizer (10-30-10) to the planting bed at 3-month intervals. Yellow, sickly leaves indicate deficiencies in minor elements, usually iron, in poor soils; apply minor element fertilizer to the planting bed at 3-month intervals.

DISADVANTAGES
: Disfiguring rust on foliage may be a problem in wet areas.

Marantaceae
(Arrowroot Family)

The arrowroot family, a group of moderate size, contains 30 genera and about 400 species. Most arrowroots are native to Central and South America or to Caribbean islands. The species of greatest economic value is West Indian arrowroot, *Maranta arundinacea,* native to the Caribbean and the northern part of South America. This plant, which has long been cultivated for its starch, has an early claim to the term arrowroot, although others in the family and several species in other families produce starchy substances that are also known by that name. Generally, the specific starches are identified according to the part of the world in which they are cultivated. Chief among these are Queensland arrowroot, from *Canna edulis* of the **canna family;** Brazilian arrowroot, or tapioca, from *Manihot esculenta* of the **spurge family;** East Indian arrowroot, from *Curcuma angustifolia* of the **ginger family;** and Polynesian arrowroot, or pia, from *Tacca leontopetaloides* of the **tacca family.**

Several other members of the Marantaceae have some utilitarian value. The Caribbean plant cachibou, *Calathea discolor,* is the source of leaves that are plaited into the popular *arima* baskets, and in some places the widespread *C. lutea* is used for house thatch. When Columbus arrived in the Caribbean, he found the native Caribs growing topee tambou *(C. alluia)* as their principal root crop. Two other Central American species, *C. macrosepala* and *C. violacea,* are grown for their tender flowers, which are sold in the markets as foodstuffs.

The family name Marantaceae is derived from that of the genus *Maranta,* honoring a sixteenth-century Venetian botanist, Bartolomea Maranti. Two possible explanations of the term arrowroot are advanced: one, that the word is a corruption of ''Aruac root,'' the Aruacs being an Indian tribe in South America; the second, that portions of certain plants in the family are said to provide antidotes to poisons for arrow tips made by aborigines.

Calathea sp.
Ice-blue Calathea

Many calatheas are grown for their ornamental foliage, but that is not true of this spectacular beauty. This native of Brazil bears some of the most ethereal flower clusters in the whole world. This fascinating plant has lavender-tipped white flowers nestled among crystalline blue bracts. The whole assemblage looks as if it were fashioned of blown glass. Some seedlings have pure white flowers and bracts, unlike the parent plant. This new species, as yet undescribed, is one of about 150 species in the genus.

The ice-blue calathea is a relatively new introduction to Hawai'i, having been imported by plant collectors Hiroshi Tagami and Richard Hart in 1973. Ultimately, it should become very popular, both as a garden ornamental and as cut flower material. The flower heads are quite similar to those of turmeric *(Curcuma domestica)* and the **jewel of Burma (*Curcuma roscoeana*).** The chief difference is in the color. The leaves of the three plants also are somewhat similar, adding to the difficulty in distinguishing among them. The ice-blue calathea is a jungle plant, and prefers sheltered garden nooks.

Calathea, from *kalathos,* meaning a basket, refers to the fact that flower heads of some species resemble baskets of flowers.

HABIT — An upright, herbaceous, evergreen plant that grows to about 5 feet in height. Large, bright green leaves, each about 18 inches long, sprout from tops of slender green leaf stalks rising from underground rhizomes. Blossoms form throughout the year. Flower clusters, 8 inches long, appear in the leaf crown; they are composed of stiff, waxy, ice-blue bracts in which nestle lavender and white flowers. Moderate growth rate; easily transplanted.

GROWING CONDITIONS — These are very tender plants, requiring considerable protection from unfavorable conditions; too much sunlight, insufficient watering, excessive wind, and poor soil conditions all affect the quality of the foliage. Grows best in moist, shaded, protected locations. Not a plant for beach areas or exposed hillsides or ridges.

USE — Specimen plant; mass planting; container plant; tropical foliage and colorful flowers.

PROPAGATION — By root division or from seed.

INSECTS/DISEASES — None of any consequence in Hawai'i.

PRUNING — Remove dead and damaged leaves and flower stalks. If plant clump becomes unsightly, cut it back to the ground level; new leaves grow readily.

FERTILIZING — Apply general garden fertilizer (10-30-10) to the planting bed at 3-month intervals and to container plants at monthly intervals. Yellow, sickly, or brown-edged leaves may indicate deficiencies in minor elements, usually iron, in poor soils; apply minor element fertilizer to the planting bed at 3-month intervals.

DISADVANTAGES — Requires more protection, care, and maintenance than is provided in most gardens.

Calathea insignis
Rattlesnake Plant

Calathea insignis, a native of the tropical American region stretching from Mexico to Ecuador, is a favorite material for flower arrangements wherever it is available. Bunches of fresh flower heads are seen often in Central and South American markets, and the dried heads are sold in considerable quantity to North American florists. They are perennial favorites of arrangers because they retain their beauty long after the blossoms have been gathered. The popular name, rattlesnake plant, has been given to this species because little imagination is needed to hear the serpent's rattles in the dried seedpods.

Plants of this *Calathea* species are considerably larger than are most of the other cultivated arrowroots, approaching the bulk of the usual clumps of heliconias or gingers. It produces large, handsome, bright green, tropical leaves, and its protracted blooming period gives the gardener a constant and generous supply of flowers. Because it is a jungle plant, it grows best in moist, shaded situations.

Calathea, from *kalathos,* meaning a basket, refers to the basket-like flower clusters produced by several members of the family; *insignis,* meaning striking or remarkable, refers to the handsome interwoven arrangement of this species' flower bracts.

HABIT An upright, herbaceous, evergreen plant that grows to about 8 feet in height. Long slender leaves are loosely arranged along tall leaf stalks that sprout from underground rhizomes. Flowers appear mostly from early spring to late fall, but occasionally during the rest of the year; flat yellow bracts in upright clusters beside the foliage contain tiny, insignificant flowers. Moderate growth rate; easily transplanted.

GROWING CONDITIONS These are very tender plants, requiring considerable protection from unfavorable conditions; too much sunlight, insufficient watering, excessive wind, and poor soil conditions all affect the quality of the leaves. Grows best in moist, shaded, protected, jungle-like locations. Not a plant for the beach or exposed hillsides or ridges.

USE Specimen plant; mass planting; container plant; tropical foliage and colorful flower clusters; fresh and dried flower clusters are used often in flower arrangements.

PROPAGATION By root division.

INSECTS/DISEASES None of any consequence in Hawai'i.

PRUNING Remove dead leaf and flower stalks after blossoming. New foliage grows back readily.

FERTILIZING Apply general garden fertilizer (10-30-10) to the planting bed at 3-month intervals. Yellow, sickly leaves indicate deficiencies in minor elements, usually iron, in poor soils; apply minor element fertilizer to the planting bed at 3-month intervals.

DISADVANTAGES Established plantings may outgrow garden areas allotted to them.

Calathea makoyana
Peacock Plant

This low-growing plant of the jungle floor, one of about 150 species of *Calathea,* is a product of the Amazon basin. Quite likely many undiscovered and unrecorded *Calathea* relatives remain in that vast, relatively unexplored area. This species and others like it have been cultivated as ornamentals for many years. The common name of the plant is based on a vague resemblance between the marks on its leaves and those on the feathers of the peacock.

The peacock plant is seen only occasionally in Hawai'i's gardens, mostly because it requires considerable attention and protection to achieve perfect growth. It is used either in masses, as a groundcover, or in protected containers, as decoration for lanais. Its colorful, variegated foliage, a mixture of subtle greens above and reds beneath, brings brightness to the shaded places where it grows best.

Calathea, from *kalathos,* meaning a basket, describes the configuration of the flower heads; *makoyana* is named for a Belgian nursery, Jacob Makoy and Company, of Liège, that did early mass-propagation of the plant, thereby making it available to the ornamental plant trade.

HABIT   A shrubby, herbaceous, evergreen plant that grows to about 3 feet in height. Round-tipped, oval, light green leaves are distinctively marked with dark green, arching spots; leaf undersides are similarly marked in light and dark reds. Insignificant flowers are not decorative, but on occasion will produce seeds. Moderate growth rate; easily transplanted.

GROWING CONDITIONS   These are very tender plants, requiring considerable protection from unfavorable conditions; too much sunlight, insufficient watering, excessive wind, and poor soil conditions all affect the quality of the leaves. Grows best in moist, shaded, protected, jungle-like locations. Not a plant for the beach or exposed hillsides or ridges.

USE   Specimen plant; mass planting; container plant; tropical foliage.

PROPAGATION   By root division.

INSECTS/DISEASES   None of any consequence in Hawai'i.

PRUNING   Remove dead and damaged leaves and flower stalks. If clump becomes unsightly, cut it back to ground level; new leaves develop readily.

FERTILIZING   Apply general garden fertilizer (10-30-10) to the planting bed at 3-month intervals and to container plants at monthly intervals. Yellow, sickly, or brown-edged leaves may indicate deficiencies in minor elements, usually iron, in poor soils; apply minor element fertilizer to the planting bed at 3-month intervals.

DISADVANTAGES   May require more care than is provided in most Island gardens.

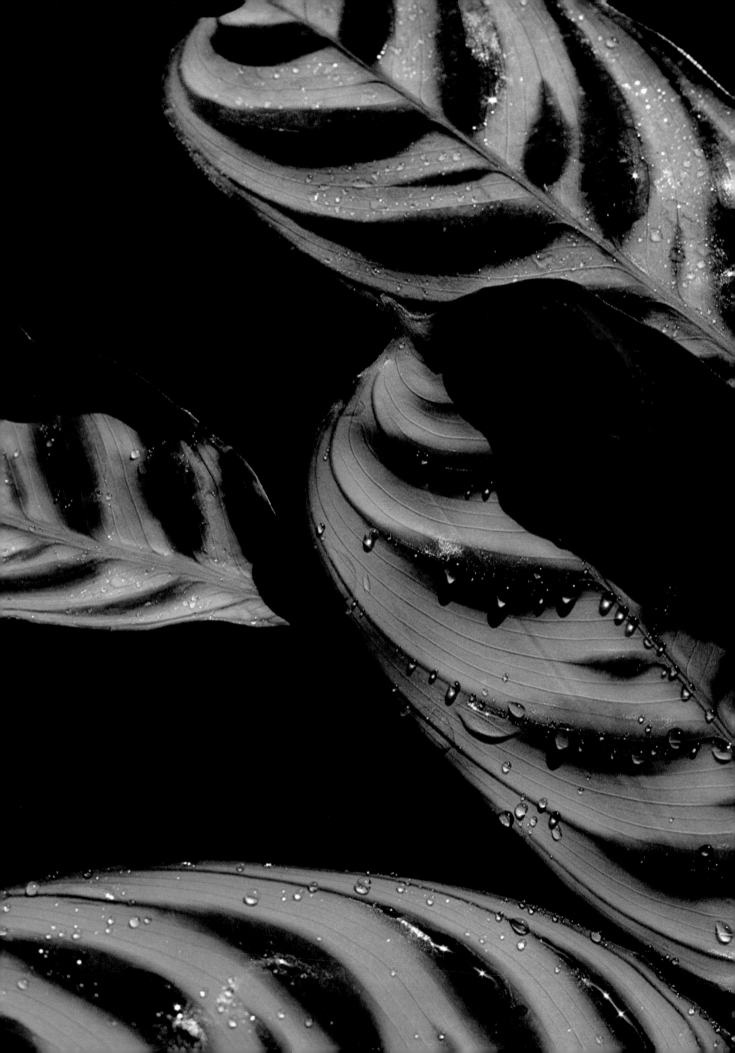

Calathea metallica
Calathea

Calathea metallica is a native of Colombia and the Amazon basin in Brazil. About 150 recognized species are included in the genus *Calathea,* all of them tropical American or Caribbean in origin. Because many of the calatheas are exquisitely ornamental, they have been plucked from their jungle habitats to serve as parent stocks for the insatiable tropical plant industry. Many of these cultivated calatheas have such similar foliage that the average plant lover is unable to distinguish one species from another. Happily, the leaves of *C. metallica* are marked rather distinctively by their jagged color patterns that resemble fish scales, in combinations of subtle greens on the upper surfaces and dramatic burgundy and silver on the lower.

The plant is very tender, requiring extreme protection from sun and wind. It is grown quite often as a potted specimen for a secluded and shaded garden lanai. Its requirement for considerable moisture in soil and air somewhat precludes its effective use as an indoor plant, especially in dry, air-conditioned rooms.

Calathea, from *kalathos,* meaning a basket, alludes to the basket-like flower clusters produced by many species in the arrowroot family; *metallica* describes the silvery undercoating of leaves.

HABIT A shrubby, ground-hugging, herbaceous, evergreen plant that grows to about 24 inches in height. Pointed, narrow, light green leaves with wavy edges are vividly marked on upper surfaces with dark green spots like fish scales, and on undersides in silvery green and red. Insignificant flowers are not decorative, but on occasion will produce seeds. Moderate growth rate; easily transplanted.

GROWING CONDITIONS These are very tender plants, requiring considerable protection from unfavorable conditions; too much sunlight, insufficient watering, excessive wind, and poor soil conditions all affect the quality of its leaves. Grows best in moist, shaded, protected, jungle-like locations. Not a plant for the beach or exposed hillsides or ridges.

USE Specimen plant; mass planting; container plant; tropical foliage.

PROPAGATION By root division.

INSECTS/DISEASES None of any consequence in Hawai‘i.

PRUNING Remove dead and damaged leaves and flower stalks. If clump becomes unsightly, cut it back to ground level; new leaves grow readily.

FERTILIZING Apply general garden fertilizer (10-30-10) to the planting bed at 3-month intervals and to container plants at monthly intervals. Yellow, sickly, or brown-edged leaves may indicate deficiencies in minor elements, usually iron, in poor soils; apply minor element fertilizer to the planting bed at 3-month intervals.

DISADVANTAGES May require more care than is provided in most Island gardens.

Calathea wiotiana
Calathea

Calathea wiotiana is another of the many ornamental low-growing plants imported from the forest floors in Brazil. Sometimes in horticultural writings it is confused with the **rattlesnake plant *(C. insignis),*** because both species have been accorded that Latin binomial at one time or another. The two species are dissimilar, however. This plant is of tender, almost evanescent, construction: the leaves are frail, approaching the thinness of tissue paper. The vividly marked and veined green upper leaves are marked in translucent rose-reds below.

A product of the dense jungles, this calathea is best grown in comparable situations. It is only occasionally used in Hawai'i's gardens because of its great dependence on humid, protected environments. For this reason it is grown fairly often in lathhouses, either as a potted plant or in beds; aficionados of tropical plants usually have at least one specimen in their collections.

Calathea, from *kalathos,* meaning a basket, refers to the basket-like construction of flower clusters produced by several members of this family; *wiotiana* is a term of uncertain origin.

HABIT A shrubby, openly arranged, herbaceous, evergreen plant that grows to about 3 feet in height. Pointed, narrow, light green leaves with wavy edges are distinctly marked on upper surfaces with variable, rounded, dark green spots and veins and on under surfaces, with similar markings in shades of pink. Insignificant flowers are not decorative, but on occasion will produce seeds. Moderate growth rate; easily transplanted.

GROWING CONDITIONS These are very tender plants, requiring considerable protection from unfavorable conditions; too much sunlight, insufficient watering, excessive wind, and poor soil conditions all affect the quality of the leaves. Grows best in moist, shaded, protected, jungle-like locations. Not a plant for the beach or exposed hillsides or ridges.

USE Specimen plant; mass planting; container plant; tropical foliage.

PROPAGATION By root division.

INSECTS/DISEASES None of any consequence in Hawai'i.

PRUNING Remove dead and damaged leaves and flower stalks. If clump becomes unsightly, cut it back to ground level; new leaves grow readily.

FERTILIZING Apply general garden fertilizer (10-30-10) to the planting bed at 3-month intervals and to container plants at monthly intervals. Yellow, sickly, or brown-edged leaves may indicate deficiencies in minor elements, usually iron, in poor soils; apply minor element fertilizer to the planting bed at 3-month intervals.

DISADVANTAGES May require more care than is provided in most Island gardens.

Maranta leuconeura
Maranta

The genus *Maranta* is small, containing only about 23 species. Among these is *M. arundinacea,* West Indian arrowroot, one of the world's most important starches. All the marantas are tropical American in origin, and several have been cultivated since people first settled there. Arrowroot starch is made from the tubers of *M. arundinacea* in a process of grinding, washing, soaking, and settling. The starch, a common thickening agent for sauces, is also a benign foodstuff for convalescents and infants.

M. leuconeura is a colorful ornamental that is very similar to the **calatheas.** It most closely resembles *C. makoyana* in that the leaf tips of both plants are definitely rounded, whereas in the other species they are quite ruffled and pointed. *M. leuconeura* can be distinguished, however, by its tricolored upper leaves, in which the contrasting greens are shot through with brilliant red veins. The plant, a creature of the jungle, dislikes sun and wind. Usually it is grown in pots for display on secluded lanais or in masses under great, protecting trees.

Maranta is named for Bartolomea Maranta, a sixteenth-century Venetian botanist; *leuconeura,* from *leucos,* meaning white, and *neuron,* meaning a nerve, describes the almost colorless smaller veins which are insignificant in comparison with the brilliant reds of the primary veins.

HABIT
A low, shrubby, herbaceous, evergreen plant that grows to about 1 foot in height. Oval leaves with rounded tips sprout in dense clusters from tuberous roots; velvety green foliage is distinctively marked, with light green coloration of the midleaf and vivid red veins; undersides of leaves are silvery rose. Insignificant flowers are not decorative but on occasion will produce seeds. Moderate growth rate; easily transplanted.

GROWING CONDITIONS
These are very tender plants, requiring considerable protection from unfavorable conditions; too much sunlight, insufficient watering, excessive wind, and poor soil conditions all affect the quality of the leaves. Grows best in moist, shaded, protected, jungle-like locations. Not a plant for the beach or exposed hillsides or ridges.

USE
Specimen plant; mass planting; container or hanging basket plant; tropical foliage.

PROPAGATION
By root division.

INSECTS/DISEASES
None of any consequence.

PRUNING
Remove dead and damaged leaves and flower stalks. If clump becomes unsightly, cut it back to ground level; new leaves grow readily.

FERTILIZING
Apply general garden fertilizer (10-30-10) to the planting bed at 3-month intervals and to container plants at monthly intervals. Yellow, sickly, or brown-edged leaves may indicate deficiencies in minor elements, usually iron, in poor soils; apply minor element fertilizer to the planting bed at 3-month intervals.

DISADVANTAGES
May require more care than is provided in most Island gardens.

Myrosma oppenheimiana cv. 'Tricolor'
Ctenanthe

This beautiful plant, a native of the Brazilian jungles, has had a checkered taxonomic career. It was first described in 1875 as *Calathea oppenheimiana;* in 1930 it was reclassified as *Ctenanthe oppenheimiana;* then, most recently (and currently), it is named *Myrosma oppenheimiana.* And it can even be found in some horticultural references as *Stromanthe porteana oppenheimiana!* Whatever its future scientific moniker may be, it has long enjoyed appreciation as an ornamental.

Generally this plant is grown in considerable quantity as a groundcover with colorful foliage. It is best used on a grand scale, in areas under large shade trees or in conjuction with plants such as the ornamental bananas, heliconias, and gingers. Like most plants in the arrowroot family, this species is quite tender, although it is sturdier and more resistant to injury than are most of its relatives.

Myrosma, from *myrum,* meaning a plant's sweet juices, and *osme,* meaning fragrance, refers to this species' scented flower clusters; *oppenheimiana* honors a nineteenth-century horticulturist.

HABIT
A low, shrubby, herbaceous, evergreen plant that grows to about 2 feet in height. Long, narrow, pointed leaves sprout in dense clusters from tuberous roots; the somewhat velvety leaves are variegated in green and gray on the upper surfaces, and with a purple cast on the lower surfaces. Flowers are insignificant and will produce seeds periodically. Moderate growth rate; easily transplanted.

GROWING CONDITIONS
These are very tender plants, requiring considerable protection from unfavorable conditions; too much sunlight, insufficient watering, excessive wind, and poor soil conditions all affect the quality of the leaves. Grows best in moist, shaded, protected, jungle-like locations. Not a plant for the beach or exposed hillsides or ridges.

USE
Specimen plant; mass planting; container plant; tropical foliage.

PROPAGATION
By root division.

INSECTS/DISEASES
None of any consequence in Hawai'i.

PRUNING
Remove dead and damaged leaves and flower stalks. If clump becomes unsightly, cut it back to ground level; new leaves grow readily.

FERTILIZING
Apply general garden fertilizer (10-30-10) to the planting bed at 3-month intervals and to container plants at monthly intervals. Yellow, sickly, or brown-edged leaves may indicate deficiencies in minor elements, usually iron, in poor soils; apply minor element fertilizer to the planting bed at 3-month intervals.

DISADVANTAGES
May require more care than is provided in most Island gardens.

Orchidaceae
(Orchid Family)

The orchid family is the largest in the whole Plant Kingdom. Consisting of 735 genera and more than 17,000 species, it easily outnumbers the next largest family, the **grasses (Gramineae),** with only 620 genera and about 10,000 species. Orchids can be found in most parts of the world. Some species are endemic only to certain alpine and tundra regions. By far the most numerous group includes the tropical and subtropical orchids, which in some areas contribute the major component of the vegetation. Hawai'i, now the adopted home of many of the tropics' ornamental species and hybrids imported from the world around, can claim only four native species: *Anoectochilus apiculatus, A. sandwicensis, Platanthera holochila,* and *Liparis hawaiiensis.* And all of these are so inconspicuous as to be uninteresting as ornamentals.

A few orchids are valuable for products other than their flowers. Most important of these species is a native of tropical America, *Vanilla planifolia,* the source of the vanilla of commerce. Vanilla extract is obtained from the fragrant dried "beans," the long, spongy seedpods of this yellow-flowered twining vine. Three or four other *Vanilla* species produce a similar flavoring, but to a lesser degree. A relative, *Orchis latifolia,* has tubers from which is made the Oriental seasoning powder, salep.

The word Orchidaceae is derived from *orchis,* which was used by Theophrastus (ca. 371–287 B.C.), the "father of botany." The name *Orchis* first appeared in Theophrastus' great botanical work, *Enquiry into Plants.* In Hawaiian the general name for all the orchids is 'okika.

Brassia gireoudiana × *B. longissima* cv. 'Edvah Loo'

Spider Orchid

This plant is a 1966 hybrid produced by crossing two wild Costa Rican species, the red-spotted, yellow-flowered *Brassia gireoudiana* and the brown-spotted, yellow-flowered *B. longissima*. The genus *Brassia* is entirely tropical American in origin and contains about 50 species. These plants have long been used in hybridization, much of which has been done in Hawai'i. Both these species and their many hybrids are valued for their fanciful and numerous ''spidery'' blossoms.

The plant is an epiphyte, growing naturally upon the limbs of trees. The flower clusters are quite large and most apparent in the garden, making this hybrid a popular ornamental. Many gardeners prefer to grow it in pots in a lathhouse, then bring them out into a viewing area during the blooming season.

Brassia is named for William Brass, an eighteenth-century English plant collector and artist who created many botanical illustrations of plants he found growing in the regions he visited; *gireoudiana* honors M. Gireoud, an obscure nineteenth-century horticulturist; *longissima,* meaning very long, describes the spider-leg petals. Edvah Loo is a Honolulu schoolgirl.

HABIT An evergreen, epiphytic plant with clusters of dark green, paired leaves rising about 10 inches above fleshy pseudobulbs. Flowers appear mostly during spring and fall months; flower clusters arch about 18 inches above the pseudobulbs; spidery, yellow-green flowers, each about 6 inches long, bloom in dense clusters along the upper stem. Somewhat dormant during summer and winter months. Moderate growth rate; easily transplanted.

GROWING CONDITIONS Naturally a tree-dweller, it must be grown as an epiphyte or in porous media, such as coarse tree fern fiber with or without charcoal chips, or in unprocessed charcoal chips alone. Requires complete drainage; may be grown in light shade or dappled sunlight. Should be watered in the mornings every 2 or 3 days in times of low humidity, and every 5 days in times of high humidity; less during dormant seasons.

USE Specimen plant; container plant; tree decoration; colorful, tropical flowers.

PROPAGATION By dividing the plant's rhizomes.

INSECTS/DISEASES To control scale, mealybugs, and thrips, apply diazinon or malathion. For spider mites, use kelthane or wettable sulfur.

PRUNING Remove old flower stalks and yellow or dead leaves.

FERTILIZING Apply slow-release, pelleted 14-14-14 fertilizer at the rate of 1 pinch per 6-inch pot every 6 months, or apply liquid foliar fertilizer (20-20-20) at monthly intervals.

DISADVANTAGES None.

Brassolaeliocattleya cv. 'Polynesian'
Cattleya Orchid

This nomenclatural tongue-twister is one of many hybrids produced by crossing species from three different genera, *Brassavola, Cattleya,* and *Laelia.* All three genera are native to parts of Central America, tropical South America, and the islands of the Caribbean. In Mexico the pseudobulbs of *Laelia speciosa* are made into paste figures used on All Saints' Day.

This cultivar, 'Polynesian,' has an unusual and important genetic feature: it is *polyploid,* that is, its chromosomes are doubled. As a result, its tissue texture is richer and heavier than that of most kinds of plants, which are not so endowed. Furthermore, this cultivar produces rich-textured flowers of extremely high and long-lasting quality. The plant was first developed in Hawai'i in the early 1960s.

Polynesian is one of Hawai'i's best winter-blooming cattleyas; most cattleyas bloom during the summer months. It may be placed on the trunks and branches of trees or be grown in pots filled with tree fern fiber.

Brassavola is named for A. M. Brassavola (1500–1555), a Venetian botanist; *Laelia* is named for a Roman vestal virgin; *Cattleya* honors William Cattley (d. 1832), British patron of botany and first successful European orchid grower, who introduced many plants into European horticulture.

HABIT An evergreen, epiphytic plant with loose clusters of dark green leaves rising about 12 inches above fleshy pseudobulbs. Flowers appear mostly from December through March; clusters of 2 to 5 large, red-violet flowers rise 18 inches above the pseudobulbs. Moderate growth rate; easily transplanted.

GROWING CONDITIONS Naturally a tree-dweller; must be grown epiphytically or in media such as coarse tree fern fiber with or without charcoal chips, or in unprocessed charcoal chips alone. Requires complete drainage; may be grown in light shade or dappled sunlight. Should be watered in the morning every 2 to 3 days in times of low humidity, and every 5 days in periods of high humidity. Flowers should be staked while blooming.

USE Specimen plant; container plant; tree decoration; colorful, tropical flowers.

PROPAGATION By dividing the plant's rhizomes.

INSECTS/DISEASES To control scale and cattleya fly, apply diazinon or malathion. For spider mites, use kelthane or wettable sulfur. For bacterial spot disease (common during extended wet periods), proper potting methods using well-aerated materials and plant sanitation are the most effective controls.

PRUNING Remove old flower stalks and yellow or dead leaves.

FERTILIZING Apply slow-release, pelleted 14-14-14 fertilizer at the rate of 1 pinch per 6-inch pot every 6 months, or apply liquid foliar fertilizer (20-20-20) at monthly intervals.

DISADVANTAGES None.

Dendrobium cv. 'Margaret Gillis'
Dendrobium Orchid

The genus *Dendrobium* is one of the larger groups in the orchid family; about 1,400 species from tropical Asia, Australia, and Polynesia are recognized. Untold numbers of hybrids have been produced throughout the world by orchid growers, and for decades Hawai'i has been one of the major centers of these hybridizations. Several *Dendrobium* species have practical uses: the flower stems of *D. crumeatum* and *D. utile* are worked into basketware in Malaysia and the Philippines; several other species are made into medicinal tonics and poulticing agents in Asia. *D. salaccense* has aromatic leaves that Malayan women wear in their hair; its leaves also add a delicate, perfumed aroma to cooked rice.

The dendrobiums rank among the most popular of Hawai'i's flowers; almost every garden displays several potted specimens. Generally the plants are grown in pots filled with tree fern fiber or with unprocessed charcoal chips. They take easily to trees. Their small size, indifference to neglect, and long blooming period make them ideal selections for small garden spaces and protected apartment lanais.

Dendrobium, from *dendron*, meaning tree, and *bios*, meaning life, refers to their epiphytic habit. This horticultural cultivar, 'Margaret Gillis,' is the result of crossing *Dendrobium* cv. 'Lady Constance' with *Dendrobium* cv. 'Maui Beauty.' This cross was made in Hawai'i.

HABIT An epiphytic plant, upright in habit, with stiff central leaf stem and many short, stiff, leathery leaves in a dense, alternate arrangement; may reach 2 to 3 feet or more in height. Flowers mostly during winter, spring, and summer; several blossoms are borne in a loose arrangement at the ends of long, slender flower stems; each rich magenta flower is about 2 inches in diameter. Fast growth rate; easily transplanted.

GROWING CONDITIONS Naturally a tree-dweller; must be grown as an epiphyte or in media such as coarse tree fern fiber with or without charcoal chips, or in unprocessed charcoal chips alone. Requires complete drainage; may be grown in light shade or considerable sunlight. Should be watered in the morning every 2 to 3 days in times of low humidity, and every 5 days in periods of high humidity. A quite adaptable plant.

USE Specimen plant; container plant; tree decoration; colorful, tropical flowers.

PROPAGATION By dividing the plant's rhizomes.

INSECTS/DISEASES To control orchid weevils, apply carbaryl. For scale, mealybugs, and thrips, apply diazinon or malathion. For spider mites, use kelthane or wettable sulfur. For black rot fungus disease (prevalent during new-growth periods), withhold water.

PRUNING Remove old flower stalks and yellow or dead leaves.

FERTILIZING Apply slow-release, pelleted 14-14-14 fertilizer at the rate of 1 pinch per 6-inch pot every 6 months, or apply liquid foliar fertilizer (20-20-20) at monthly intervals.

DISADVANTAGES None.

238

Epidendrum spp. hybrids
Reed Epidendrums

The epidendrums are tropical American in origin; about 400 species are known at present. A few have some practical uses: *Epidendrum bifidum,* from tropical South America, is employed by natives of that region for treatment of tapeworm and similar intestinal complaints; *E. cochleatum,* from Central and tropical South America and the Caribbean, has pseudobulbs which are processed into a mucilage. Many ornamental hybrids have been produced from the best of the collected species. Epidendrums are closely related to the genus **Cattleya.**

In Hawai'i, the "epis," as they are known, are almost always planted in beds filled with coarse crushed rock or tree fern fiber, even though their natural habitat is the trunks and branches of trees. Sometimes they are called reed epidendrums, because of the appearance of their rounded, thin stems and leaves. The plants are vertical in habit. Color variations abound, the most favorite probably being in the lavender range, but whites, pinks, yellows, oranges, and red-oranges are also popular. Of all orchids in Hawai'i, epis are most often seen in mass plantings or on rock walls.

Epidendrum, from *epi,* meaning upon, and *dendron,* meaning a tree, refers to the epiphytic habit of most members of the genus.

HABIT A vertical, clumping, epiphytic orchid with a stiff, central leaf stem that holds many stiff, flat, bright green leaves in an open, alternate arrangement; plant may reach 3 to 4 feet or more in height; new rooted shoots appear along flower stems of older plantings. Flowers appear mostly during spring and summer; many blossoms are borne in a dense cluster at the tips of long, slender flower stems; tiny, half-inch flowers, quite frilly, are white, pink, red-orange, yellow, purple, or lavender. Fast growth rate; easily transplanted.

GROWING CONDITIONS Dwelling naturally in trees or on faces of rocks, these plants, when grown elsewhere, must be set in highly aerated substrates, such as coarse tree fern fiber, crushed rock, or charcoal chips. Require complete drainage; may be grown in light shade or full sunlight. Should be watered in the morning every 2 or 3 days in times of low humidity, and every 5 days in periods of high humidity. Very adaptable.

USE Specimen plant; mass planting; container plant; tree or wall decoration; colorful, tropical flowers.

PROPAGATION From shoots that develop roots on the flower stalks of the parent plant.

INSECTS/DISEASES To control scale, mealybugs, and thrips, apply diazinon or malathion. For spider mites, use kelthane or wettable sulfur.

PRUNING Remove old flower stalks and yellow or dead leaves.

FERTILIZING Apply slow-release, pelleted 14-14-14 fertilizer at the rate of 1 pinch per 6-inch pot every 6 months, or apply liquid foliar fertilizer (20-20-20) at monthly intervals.

DISADVANTAGES None.

240

Ludisia discolor
Jewel Orchid

The genus *Ludisia* consists of only one species, *L. discolor.* Several hybrids have been produced, however, generally by crossing this plant with closely related species in other genera. Probably the most common hybrids are those made between *Ludisia* and *Anoectochilus* spp.: the resulting progeny are known as *Anoectomaria* hort. (The genus *Anoectochilus* is of added interest to Hawaiian plant lovers in that it contains two of the four native Hawaiian orchids, *A. apiculatus* and *A. sandwicensis.*) Similar jewel orchid hybrids have been made between *Ludisia discolor* and species of *Macodes,* producing hybrids called *Macomaria* hort. Results of all these various crosses are known collectively as jewel orchids.

Ludisia discolor is found growing naturally throughout much of Indochina and the Malay Peninsula, thriving in moist, protected forest areas, near streams and among rock outcroppings. Several foliar varieties are known, most of them extremely beautiful. If the plant is given adequate care, it will provide the gardener with year-round beauty, both in its colorful velvety foliage and in its lovely fragrant white flowers.

Possibly *Ludisia* may be derived from a Malay name for the plant, duan lo, which has been translated as Low's leaf, honoring Hugh Low (1824–1905), a British plant collector and nurseryman who spent much time in Malaysia and Indonesia; *discolor,* meaning of different colors, refers to the several leaf forms with their assorted color variations.

HABIT A terrestrial plant that acts somewhat like an epiphyte, growing in light decayed material of the forest floor. Produces a few leaves in an informal whorl; subsequent root offshoots produce similar foliage, making a low, velvety groundcover; plants grow to about 1 foot in height. Velvety foliage is dark wine-red and green. In the fall and winter, small clusters of fragrant white flowers appear on long, slim stems above the foliage mass. Fast growth rate; easily transplanted.

GROWING CONDITIONS Naturally a plant of the damp forest floor; grows best in planting media extremely high in humus; complete aeration and drainage are required. The plant is extremely tender; its foliage must be protected from sunburn. Should be watered daily in the morning hours. Requires high humidity.

USE Specimen plant; mass planting; container plant; colorful foliage and fragrant flowers.

PROPAGATION By dividing the plant's rhizomes.

INSECTS/DISEASES To control scale, mealybugs, and thrips, apply weak solutions of diazinon. For spider mites, use weak solutions of kelthane or wettable sulfur.

PRUNING Remove old flower stalks and yellow or dead leaves.

FERTILIZING Apply slow-release, pelleted 14-14-14 fertilizer at the rate of 1 pinch per 6-inch pot every 6 months, or apply liquid foliar fertilizer (20-20-20) at monthly intervals.

DISADVANTAGES Delicate compared with other orchids, requiring environments of high humidity.

Oncidium cv. 'Valverde'
Dancing Doll Orchid

The oncidiums are native to Florida, the West Indies, and temperate South America. About 350 species comprise the genus. The oncidiums are called dancing doll orchids because each flower cluster appears to be an entire corps de ballet attired in frilly tutus. This orange-flowered cultivar is a descendant of no fewer than five different species from the Caribbean: *Oncidium desertorum, O. henekenii, O. triquetrum, O. urophyllum,* and *O. variegatum.* Its immediate parents are the hybrids *Oncidium* cv. 'Red Belt' and *Oncidium* cv. 'Yellow Jacket.' Probably the most common species for landscaping is the yellow-and-brown-flowered *O. altissimum,* from tropical America. Most often it is grown on trees or in pots in Hawaiian gardens, giving a forsythia-like show of yellow during the spring months.

The numerous kinds of oncidiums are popular garden plants in the Islands. Naturally tree dwellers, they are so placed on the branches that the long flower stems hang in graceful cascades. They are simple to grow and easy to maintain, and become naturalized easily in the Hawaiian landscape.

Oncidium, from *onkos,* meaning a tumor, refers to the flowers' warted lips; the horticultural term, 'Valverde,' or green valley, was given by the hybridizers.

HABIT	A small evergreen epiphytic plant, about 6 inches high, with dense clusters of light green leaves rising about 8 inches above fleshy pseudobulbs. Flowers appear mostly between late fall and early summer; dense clusters of many orange-colored, ruffled flowers, each about ½ inch in diameter, crowd the 8-inch, arching flower stems. Fast growth rate; easily transplanted.
GROWING CONDITIONS	Naturally a tree dweller; must be grown as an epiphyte, or in media such as coarse tree fern fiber with or without charcoal chips, or in unprocessed charcoal chips alone. Requires complete drainage; may be grown in light shade or dappled sunlight. Should be watered in the morning every 2 to 3 days in times of low humidity, and every 5 days in periods of high humidity. Do not overwater.
USE	Specimen plant; container plant; tree decoration; colorful, tropical flowers.
PROPAGATION	By dividing the plant's rhizomes.
INSECTS/DISEASES	To control scale, mealybugs, and thrips, apply diazinon or malathion.
PRUNING	Remove old flower stalks and yellow or dead leaves.
FERTILIZING	Apply slow-release, pelleted 14-14-14 fertilizer at the rate of 1 pinch per 6-inch pot every 6 months, or apply liquid foliar fertilizer (20-20-20) at monthly intervals.
DISADVANTAGES	None.

Paphiopedilum wolterianum
Lady Slipper Orchid

The paphiopedilums are native to the forest floors of the Malay archipelago, Indonesia, and islands ranging eastward to the Solomons. About 50 species are represented in the genus. A very closely allied orchid group, North Temperate Zone in origin, is the genus *Cypripedium.* Two cypripediums native to North America are commonly known as moccasin flowers. One, *C. reginae,* is the state flower of Minnesota; the dried rhizomes of the other, *C. pubescens,* have been used both as a nerve stimulant and an antispasmodic.

The paphiopedilums and cypripediums are similar in appearance and are cultivated in much the same way. Generally they are presented as potted specimens, although, with extreme care, they may be grown in beds with high humus content, in shaded and protected locations. They are not good house plants, but blooming potted specimens may be brought indoors for a few hours' display. Often the flowers are used in corsages, especially during winter months—their blooming period, when most other orchids are dormant.

The genus name *Paphiopedilum* combines the name of a city in Cyprus (Paphos) sacred to Venus, and the Latin word *pes,* meaning foot (hence, slipper); *wolterianium* commemorates Paul Wolter (b. 1861), an orchid grower of Magdeburg, Germany.

HABIT
A terrestrial orchid that produces sturdy, flattened, dark green leaves arranged equally on two sides of the plant. Flowers appear mostly during winter and spring; a single flower stem sprouts from the center of a leaf cluster to produce a single purple and green flower, about 3 inches in diameter. Slow growth rate; can be transplanted with care.

GROWING CONDITIONS
Naturally a forest-floor plant; must be grown in a rich humus mixture, such as fir bark or leafmold. Requires complete drainage; may be grown in light shade or dappled sunlight. Should be watered in the morning every 2 to 3 days in times of low humidity, and every 5 days in periods of high humidity.

USE
Specimen plant; container plant; colorful, tropical flowers.

PROPAGATION
By dividing the plant's rhizomes.

INSECTS/DISEASES
To control scale, mealybugs, and thrips, apply diazinon or malathion. For spider mites, use kelthane or wettable sulfur.

PRUNING
Remove old flower stalks and yellow or dead leaves.

FERTILIZING
Apply slow-release, pelleted 14-14-14 fertilizer at the rate of 1 pinch per 6-inch pot every 6 months, or apply liquid foliar fertilizer (20-20-20) at monthly intervals.

DISADVANTAGES
None.

Phalaenopsis cv. 'Lavender Lady'
Butterfly Orchid, Mariposa

The butterfly orchids are native primarily to Southeast Asia and Indonesia, although a few species come from southern China, the Philippines, and Australia. About 35 species of *Phalaenopsis* are recognized. They are almost exclusively ornamental and have been the subjects of considerable hybridization by the world's orchid growers. In general, hybridization has produced cultivars that will bear much larger and more beautiful flowers than are borne by the natural species. *Phalaenopsis* cv. 'Lavender Lady,' created in California in 1967, is a cross between *Phalaenopsis* cv. 'Best Girl' and *Phalaenopsis* cv. 'Zada,' each in itself a hybrid.

The plants grow naturally on tree trunks and branches in the jungle forests, but grow equally well when established on lava rock walls or tree ferns. They hang gracefully, showing cascades of color during the cool-weather blooming season. Often they are cultivated in a lathhouse, in pots filled with tree fern fiber, then are brought out for display when they bloom.

Phalaenopsis, from *phalaina*, meaning a moth, and *opsis*, meaning appearance, describes the flowers' resemblance to butterflies. Filipinos know them as mariposa, the Spanish name for butterfly.

HABIT | An epiphytic plant with a few large, strap-shaped leaves arranged more or less evenly on the two sides of the plant; leaves grow to about 6 inches in height. Many flowers appear during winter and spring; long slender arching stems bear several pink butterfly-like flowers, each about 4 inches in diameter; extensive plantings suggest the appearance of cascades. Slow growth rate; easily transplanted.

GROWING CONDITIONS | Naturally a tree-dweller; must be grown as an epiphyte or in media such as coarse tree fern fiber with or without charcoal chips. Requires complete drainage; may be grown in light shade or dappled sunlight. Should be watered in the morning every 2 to 3 days in times of low humidity, and every 5 days in periods of high humidity. Will not withstand drought.

USE | Specimen plant; container plant; tree decoration; colorful, tropical flowers.

PROPAGATION | New hybrids are grown from seed, but the process is difficult for the home gardener. Generally propagated by offshoots which appear on old flower stems.

INSECTS/DISEASES | To control orchid weevils, apply carbaryl. For scale, mealybugs, and thrips, apply diazinon or malathion. For spider mites, use kelthane or wettable sulfur.

PRUNING | Remove yellow or dead leaves. Do not remove flower stalks because new blooms appear on old stalks year after year.

FERTILIZING | Apply slow-release, pelleted 14-14-14 fertilizer at the rate of 1 pinch per 6-inch pot every 6 months, or apply liquid foliar fertilizer (20-20-20) at monthly intervals.

DISADVANTAGES | None.

Spathoglottis plicata var. *alba*
Ground Orchid

The *Spathoglottis* species are terrestrial orchids native to China, Indo-Malaysia, and islands eastward to Australia, the Solomons, and Fiji. They are common in the open fields and along the banks of streams in their native habitats. About 46 species comprise the genus. Several of the species are grown for their ornamental value, probably the most common being the purple-flowered *S. plicata*. This plant now grows wild on Hawai'i's open lower ridges and along the roadsides, having escaped from cultivation. Its white-flowered variety, *alba,* although less well known and less widely dispersed, is beautifully ornamental, and makes a bright show among grasses and roadside weeds when it is found in the wild.

These plants are best displayed in masses in the garden. The leaves are large, bright green, and handsomely pleated, giving considerable beauty in themselves during times when blossoms are not on display. When blossoms do appear, the effect is that of beds of spring flowers such as are seen in colder climates. Beds containing large numbers of these orchids bring welcome masses of contrasting textures and colors to Island gardens during their long blooming season.

Spathoglottis, from *spathe,* meaning spathe, and *glottis,* meaning tongue, describes the flower parts; *plicata,* meaning pleated, refers to the handsome leaves; *alba* means white.

HABIT
A terrestrial orchid often found growing naturally in grassy fields and on hillsides; the plant produces arching, pleated and pointed, bright green leaves from a central root, each leaf about 2 feet long. Blossoms appear during much of the year; several white flowers, often flushed with pink, each 1 inch in diameter, are borne on a slender, foot-long stem that sprouts from the central leaf cluster. Produces seeds readily; reseeds itself. Fast growth rate; easily transplanted.

GROWING CONDITIONS
Grows best in open, sunny areas in humus-rich, well-watered, well-drained soil. A very adaptable plant; it is one of the easiest orchids to cultivate.

USE
Specimen plant; mass planting; container plant; tropical foliage and colorful flowers.

PROPAGATION
Easily propagated by division of rhizomes. Water immediately after planting, but do not overwater in succeeding days.

INSECTS/DISEASES
To control scale, mealybugs, and thrips, apply diazinon or malathion. For spider mites, use kelthane or wettable sulfur.

PRUNING
Remove old flower stalks and yellow or dead leaves.

FERTILIZING
Apply slow-release, pelleted 14-14-14 fertilizer at the rate of 1 pinch per 6-inch pot every 6 months, or apply liquid foliar fertilizer (20-20-20) at monthly intervals.

DISADVANTAGES
None.

250

Vanda cv. 'Patricia Low'
Vanda Orchid

The vandas originated in China, Indo-Malaysia, and some islands of the western Pacific. They are epiphytic orchids, growing naturally on trees or rock outcroppings. They tend to be rather upright and climbing, often covering a considerable area. One species, *Vanda hookeriana,* scrambles over bushes in its native Malaysian swamps; known as the bone plant to Malays, sometimes parts of this orchid are applied as a hot poultice in treatment of afflictions such as arthritis and rheumatism. The root of *V. tessalata,* from India and Sri Lanka, is employed in similar fashion by Hindu physicians.

Vandas are most valuable, of course, as producers of ornamental flowers. *Vanda* cv. 'Miss Joaquim,' a hybrid of the Malayan bone plant and *V. teres,* is probably the most widely grown of all orchids. It is the source of flowers for Hawai'i's commonest orchid leis. Hawai'i is world-renowned for hybridization of vandas, and varieties are produced commercially for both local use and export to the mainland. *Vanda* cv. 'Patricia Low' (shown here) is prized for the beauty and longevity of its flowers, and for its adaptability.

Vanda is a Sanskrit name long used for a specific type of orchid. The hybrid 'Patricia Low' was developed in Singapore in 1961; it is a cross between two other hybrids, *Vanda* cv. 'Josephine van Brero' and *Vanda* cv. 'Jennie Hashimoto.' Patricia Low is a former resident of Singapore.

HABIT An epiphytic plant, vertical in habit, with a stiff, central leaf stem that holds many long, stiff, leathery leaves in a dense, alternate arrangement; plant may reach 3 to 4 feet or more in height; taller plants produce thin aerial roots which help to hold them upright. Blossoms appear during winter, spring, and summer; several are borne in a compact cluster from a stem that juts out from the leaf mass; each rust-pink flower is about 4 inches in diameter. Moderate growth rate; easily transplanted.

GROWING CONDITIONS Naturally a tree dweller; grows best as an epiphyte or in media such as coarse tree fern fiber with or without charcoal chips, or in unprocessed charcoal chips alone. Requires complete drainage; may be grown in light shade or full sunlight. Water in the morning every 2 to 3 days if humidity is low, every 5 days if it is high.

USE Specimen plant; container plant; tree decoration; colorful, tropical flowers.

PROPAGATION Remove top of plant with aerial roots attached and set on tree fern fiber. Parent plant will produce new growth below the cut end.

INSECTS/DISEASES To control orchid weevils, apply carbaryl. For scale, mealybugs, and thrips, use diazinon or malathion. For spider mites, use kelthane or wettable sulfur. For black rot fungus disease (prevalent in overwet conditions), apply captan or maneb.

PRUNING Remove old flower stalks and yellow or dead leaves.

FERTILIZING Apply slow-release, pelleted 14-14-14 fertilizer at the rate of 1 pinch per 6-inch pot every 6 months, or apply liquid foliar fertilizer (20-20-20) at monthly intervals.

DISADVANTAGES None.

Appendix 1
Insect Pests and Plant Diseases

Sucking Insects

SCALE INSECTS There are several kinds of scale insects; the species most commonly found in Hawai'i are the green scale, white armored scale, red wax scale, and barnacle scale. These insects live by sucking nutrients from the leaves, stems, and branches of the many susceptible plants. The actual insect lives protected in its familiar round or flattened, disklike cover attached securely to the plant tissue. These covers may be brown, black, white, green, or pink in color.

Control scale by spraying malathion in 2 applications, 14 days apart, on the upper and lower surfaces of the foliage and on the stems and branches. Natural biological control is provided by ladybird beetles, which feed on these insects.

Scale insects exude a honeydew that attracts ants, which then become a pest, and also provides nourishment for sooty mold (a fungus). To control ants, apply malathion or diazinon to the plant's base. To eradicate or prevent sooty mold, the scale insects must be controlled.

MEALYBUGS Mealybugs are closely related to the scale insects. They are curiously shaped, white-coated insects that resemble slow-moving pieces of waxy cotton lint. They are generally found in colonies on the surfaces of the leaves, stems, branches, and roots of many Island plants. Several generations are produced during the course of a year. They feed by sucking nutrients from the plant tissue.

Control mealybugs by applying diazinon or malathion sprays in 2 treatments, 14 days apart, to the upper and lower surfaces of the leaves and to the stems and branches. Drench the surrounding soil with the same insecticides if mealybugs have attacked the plant's roots.

Mealybugs exude a honeydew that attracts ants, which then become a pest, and also provides nourishment for sooty mold (a fungus). To control ants, apply malathion or diazinon to the plant's base. To eradicate or prevent sooty mold, the mealybugs must be controlled.

THRIPS Several species of thrips are found in Hawai'i; some of them attack only certain plant species. They are particularly attracted to gardenia flowers and Chinese banyans. Thrips are small, slender, shiny black insects that appear in considerable numbers. A characteristic feature is that they raise their tails in a menacing attitude as if to sting (which they are not capable of doing). These insects live by sucking sap from the plant's flowers, leaves, and stems. Damage to the plant consists of malformed or unopened flower buds and lackluster, curled leaves. The foliage often turns silvery from the rasping, sucking feeding action of the insects.

Control thrips by applying diazinon or malathion in 2 treatments, 14 days apart, to the upper and lower surfaces of the flowers and leaves, and to the branches.

APHIDS These are small, soft, dull-colored insects that appear in dense clusters along the stems, leaves, buds, and flowers of many plants. Sometimes plant parts are almost completely covered by multitudes of these insects. They feed by sucking out the sap, which deforms buds, pits fruits and vegetables, and discolors and curls foliage. Numerous generations are produced throughout the year. Some adults develop wings which allow them to fly to new feeding areas.

Control aphids by applying dimethoate, diazinon, or malathion spray in 2 treatments, 14 days apart, to the upper and lower surfaces of the foliage, to flowers, and to the stems and branches. Aphids exude a honeydew that provides nourishment for sooty mold (a fungus), which is controlled by eradicating the aphids.

Ants use the aphids as cattle, milking the honeydew from the aphids' bodies. Control ants by applying malathion or diazinon to the plant's base. Aphids are carriers of several virus diseases, such as the dread papaya mosaic virus. Eradication of aphids is one of several control measures for virus diseases.

SOUTHERN GREEN STINKBUGS — These are the unwelcome, shield-shaped, green or bronze-green insects that produce a highly offensive odor when pinched or stepped on. They live by sucking sap from leaves, stems and fruits of many plant species, causing deformed leaves, stems, and fruits. Often the plant tissue becomes infected at the hole where the insect has fed, causing a rotten fissure to develop.

Control this pest by applying diazinon or malathion sprays in 2 treatments, 14 days apart, to the upper and lower surfaces of the foliage and to the branches and fruit. Two beneficial insects have been introduced in Hawai'i to prey (as parasites) on the southern green stinkbug—an effective biological control.

SPIDER MITES — These tiny creatures are true spiders that live in multitudinous colonies on the flowers, leaves, and stems of many plants. Spider mites are so small that they can usually be seen only under a magnifying glass. Damage to the plant is the visual evidence that the mites are present. These creatures suck the sap from plant tissues, causing discolored and deformed leaves and flowers. Often the foliage turns a listless yellow and the leaves curl; fruits are left with a russet, scaly scar where the spiders have fed.

Control spider mites by applying wettable sulfur or kelthane sprays in 2 treatments, 14 days apart, to the upper and lower surfaces of leaves and to stems, branches, flowers, and fruits. Damage to fruit may be prevented by spraying the plant just after the main flowering period when the fruit begins to form. To protect the leaves from attack, the chemical should be applied during the time when the new foliage appears.

Chewing Insects

CHINESE ROSE BEETLES — These beetles are among the most voracious and destructive chewing pests in Hawai'i, feeding on many species of plants. They are more commonly found in the hot, dry Island areas than in the cooler, moister sections. The tan-colored, oval-shaped adult beetles are about ¼ inch in length. They fly readily, hide during the day under dry plant litter, and emerge at dusk to feed until nearly midnight. They will not feed during daylight hours or at night if plantings are well lighted. The insect feeds on central portions of a plant's leaves (not along the edges) and may cause extremely heavy damage, making lacework of the foliage. Tender, large-leafed plants, such as the acalyphas, gingers, heliconias, mountain apple, grapes, avocados, and false kamani, are particularly susceptible to Chinese rose beetle attack.

Control this pest by applying one of the residual insecticides, such as carbaryl, in 2 treatments, 14 days apart, to the upper and lower surfaces of the foliage. Control is aided by keeping the garden area clear of all fallen leaves, debris, and compost, so that the insects cannot hide and breed among the ground litter. It is also helpful to light susceptible plants artificially during the feeding hours.

GRASSHOPPERS   Several grasshopper species, easily identified by their unique shapes, are found in Hawaiian gardens. Like Chinese rose beetles, these pests inflict severe damage on many garden plants. Damage from grasshoppers is easily distinguished from that of rose beetles, for grasshoppers begin chewing at the outer margin of the leaf rather than at the center. They are seldom seen feeding, as they fly away from the leaves after taking their fill.

Grasshoppers may be controlled by applying one of the residual insecticides, such as carbaryl or dursban, in 2 treatments, 14 days apart, to the upper and lower surfaces of the foliage.

CATERPILLARS   Several destructive moth and butterfly caterpillars are common in Hawaiian gardens. Some of these are the large oleander hawk moth, citrus swallowtail butterfly, monarch butterfly, and various smaller moths. The caterpillars may devour much of a plant's foliage overnight, and also may badly damage some flowers and flower buds (unopened flower buds of hibiscus, crown flowers, and geraniums are particularly susceptible).

To control the various moth and butterfly caterpillars, apply one of the residual insecticides, such as carbaryl, in 2 treatments, 14 days apart, to the upper and lower surfaces of the leaves and to the buds and flowers.

CATTLEYA FLY   The cattleya fly is actually a small black wasp. It is about 3/16 inch long. It lays its eggs in the fleshy stem (pseudobulb) of the orchid plant. The larvae feed on the tissues of the pseudobulb, causing it to become swollen. The larvae pupate and the flies emerge through a small hole in the side of the pseudobulb. The result is a hollow, unhealthy orchid stem.

Control cattleya fly by applying diazinon or malathion in 2 treatments, 14 days apart, to the pseudobulbs and foliage. As a preventive measure, apply diazinon or malathion at monthly intervals.

ORCHID WEEVILS   Two insects, quite similar in appearance, are known as orchid weevils. One is about 1/4 inch long; the other, called the lesser orchid weevil, is about half as long. Both are black, snouted weevils. Using their snouts, they drill holes in the sides of the orchid pseudobulbs and deposit their eggs within the plant tissue. The larvae feed on the interior tissues of the plant, and adults emerge from holes in the sides of the pseudobulbs. This attack causes the plant parts to turn black and die.

Control orchid weevils by applying carbaryl in 2 treatments, 14 days apart, to the upper and lower surfaces of the pesudobulbs and the foliage. As a preventive measure, apply carbaryl at monthly intervals.

Other Infestations and Diseases

SOOTY MOLD   This fungus disease is characterized by a black, sooty covering on various parts of many plant species. It lives on the honeydew secreted by scale insects, mealybugs, aphids, and whiteflies. Excessive growth of sooty mold is deleterious to the plant in that it interferes with photosynthesis; the leaves weaken and ultimately shrivel. To control sooty mold, the insects that produce the honeydew must be eradicated. Lack of honeydew causes the fungus to roll up like sunburned skin and drop from the plant.

POWDERY MILDEW   This disease is easily identified by its talcum-like appearance on the foliage, stems, and flower buds of several plant species. It is especially common on roses, acalyphas, and hydrangeas, but also afflicts many vegetables and other soft, herbaceous plants. It is most often seen during periods of damp weather.

Control powdery mildew by 2 applications of powdered or wettable sulfur or liquid mildew fungicide, 14 days apart, on the upper and lower surfaces of the leaves, stems, and flower buds.

NEMATODES More than 25,000 species of these tiny eel worms are known throughout the world. Some of them attack plant roots, sucking nutrients from the tissues. Damage is so severe, often resulting in greatly deformed, almost cancer-like root development, that the roots are unable to carry out their nutrient-gathering functions for the growing plant. Evidence of nematode infestation may be seen by inspecting visually the plant's roots. An above-ground symptom is severe wilting of plants during the heat of the day, even when the surrounding soil is moist, because the roots are unable to absorb enough moisture to keep the plant supplied during the warm, dry period. The presence of nematodes is also suspected when patches of unhealthy plants appear in otherwise healthy plantings.

Nematode attack can be prevented by selecting resistant varieties (e.g., local Hawaiian tomatoes are bred for resistance to nematodes). Soil nematodes can be eliminated only by fumigating the planting soil where they are present; this process should be undertaken only by experienced fumigators authorized by the Hawaii State Department of Agriculture. For potted plants, use sterilized potting soil. Do not put into the soil new plants that show any sign of attack by nematodes. The healthy tops of diseased plants may be used for propagation and their roots dug up and destroyed. All plant material that shows evidence of nematode attack should be destroyed.

RUST Rust disease is characterized by the presence of numerous yellowish or orange clusters of spores, usually on the undersides of a plant's leaves. Although many kinds of plants are susceptible, the disease is particularly prevalent on orchids, cannas, and spider lilies. The spores are borne through the air from infected plants to healthy ones, where they need only a small amount of moisture to begin to grow.

The best control measures are to remove all infected foliage within the garden area, and to avoid wetting the foliage of susceptible plants, if at all possible. Plants grown in well-ventilated areas are less likely to be troubled by rust than those that are placed where the air is stagnant.

BLACK ROT The first signs of black rot are the appearance of small watery spots under the surface of orchid leaves and decaying of roots. In a very short time the affected plant tissues turn black and die. The disease is caused by a fungus that multiplies rapidly during periods of hot, humid weather and an overabundance of water.

Control black rot by applying captan or benlate in a single treatment at the first sign of the condition, and repeat when necessary.

BACTERIAL SPOT The first signs of this disease are very small discolored spots at the tips of the plant's leaves. As the disease progresses, the spots enlarge and form dark crater-like depressions on the leaves. The disease is spread from plant to plant by water-borne spores in moist environments.

To control bacterial spot disease, remove all infected foliage within the garden area, and also avoid wetting the foliage of susceptible plants, if at all possible. Plants grown in well-ventilated areas are less troubled by bacterial spot disease than are those placed where the air is stagnant and moist.

258

Appendix 2
Plant Propagation

CUTTINGS Dip the cut ends of 4-inch-long stem cuttings, with half the leaves removed, in growth regulator, and plant in vermiculite or perlite. Place container in a partially shaded location and keep lightly moist. When cuttings are well rooted, transplant to a permanent garden site.

SEEDS Remove ripe seeds from pods and fleshy seedcoats and for best results, plant immediately. Sprinkle soil lightly over seeds in a bed of well-drained potting soil; cover seeds to a depth approximating the dimension of their diameters; tiny, dustlike seeds are generally scattered on the soil surface and left uncovered. Keep the seedbed lightly moist. Seeds usually germinate over a period of several weeks, although those of a few species may take months to sprout. Transplant seedlings to permanent garden locations when they reach 2 or more inches in height and have developed at least a half dozen mature leaves. For soft, translucent seeds, such as those of the arum family, sow directly on the top of shredded tree fern fiber or sphagnum moss, and keep lightly moist in partial shade. Humidity may be maintained by placing a pane of glass over the seedbed container.

ROOT DIVISION Clumping plants are easily propagated by root division. Remove rooted offshoots from the main root mass by severing with a sharp knife or by lifting the entire root mass and dividing the clump into several sections by cutting or pulling apart. Transplant divisions to protected garden locations and keep constantly moist. Most divided plants adapt readily to new garden sites. Do not fertilize new plants until they show vigorous new foliar growth.

GRAFTING Plants are often propagated by grafting in order to preserve established varieties and to obtain strong-rooted plants from otherwise weak-rooted cutting material. Some plants, such as the many hibiscus hybrids, have inherently weak roots; grafting allows the weak-rooted but desirable hybrids to be established on stronger and more vigorous rooted stock. The process may be done in several ways, the most common of which is the "side graft," which is described here.

In this process a cutting (scion) of a desired plant is embedded into the stem of a strong-rooted relative in such a way that the growing surfaces of both scion and rootstock are pressed together to allow the tissues to heal and grow together. Generally, the rootstock is about ¾ inch in diameter and the scion, slightly smaller. Cut the base of a 3-inch length of the scion into a two-sided sharp wedge and insert it into a diagonal gash cut about a quarter of the way through the diameter of the stock. Seal the junction with plastic tape or raffia covered with melted paraffin. The graft should begin to produce new foliage within about 4 months. After the scion shows vigorous growth, prune off all rootstock growth directly above the graft, and seal the cut with a tree pruning compound. This allows the graft to be established as the sole crown of the plant. As the plant grows, keep all sucker growth removed from below the graft union so that the rootstock cannot overwhelm the graft.

AIRLAYERING Roots may be generated on the branches and trunks of established woody shrubs, such as dieffenbachia, aglaonema, ti, and the dracaenas, by airlayering. First, make two parallel cuts through the bark, completely around a ½-inch-diameter branch or stem; the cuts should be one inch apart. Remove the bark between the two cuts, laying the wood beneath bare. Apply a small amount of growth regulator to the stem section where the bark has been removed. Cover the cut surface with a handful of dampened but not overly wet sphagnum moss. Cover the dampened moss with a small rectangle of polyethylene plastic sheeting, and tie securely at both edges of the plastic so that the moss is completely wrapped and covered. Be sure that the ends of the plastic overlap at the seam so that the moistened moss is firmly contained. Roots develop along the cut surface within about 4 months' time. The roots, when well developed, can be easily seen through the transparent plastic sheeting.

When well rooted, sever the branch from the parent plant and discard the plastic cover. At the time of separation from the parent, remove about a third of the foliage from the airlayered branch to prevent wilting. Set the rooted airlayer in extremely well drained potting soil, and stake it firmly upright to hold it in place. Set the potted airlayer in a partially shaded location for approximately one month, then in a sunnier area for 2 weeks before setting it out in the permanent garden location.

Index of Plant Names

262

Index